Lecture Notes in Physics

New Series m: Monographs

The Editorial Policy for Monographs

The series Lecture Notes in Physics reports new developments in physical research and teaching - quickly, informally, and at a high level. The type of material considered for publication in the New Series m includes monographs presenting original research or new angles in a classical field. The timeliness of a manuscript is more important than its form, which may be preliminary or tentative. Manuscripts should be reasonably self-contained. They will often present not only results of the author(s) but also related work by other people and will provide sufficient motivation, examples, and applications.
The manuscripts or a detailed description thereof should be submitted either to one of the series editors or to the managing editor. The proposal is then carefully refereed. A final decision concerning publication can often only be made on the basis of the complete manuscript, but otherwise the editors will try to make a preliminary decision as definite as they can on the basis of the available information.
Manuscripts should be no less than 100 and preferably no more than 400 pages in length. Final manuscripts should preferably be in English, or possibly in French or German. They should include a table of contents and an informative introduction accessible also to readers not particularly familiar with the topic treated. Authors are free to use the material in other publications. However, if extensive use is made elsewhere, the publisher should be informed.
Authors receive jointly 50 complimentary copies of their book. They are entitled to purchase further copies of their book at a reduced rate. As a rule no reprints of individual contributions can be supplied. No royalty is paid on Lecture Notes in Physics volumes. Commitment to publish is made by letter of interest rather than by signing a formal contract. Springer-Verlag secures the copyright for each volume.

The Production Process

The books are hardbound, and quality paper appropriate to the needs of the author(s) is used. Publication time is about ten weeks. More than twenty years of experience guarantee authors the best possible service. To reach the goal of rapid publication at a low price the technique of photographic reproduction from a camera-ready manuscript was chosen. This process shifts the main responsibility for the technical quality considerably from the publisher to the author. We therefore urge all authors to observe very carefully our guidelines for the preparation of camera-ready manuscripts, which we will supply on request. This applies especially to the quality of figures and halftones submitted for publication. Figures should be submitted as originals or glossy prints, as very often Xerox copies are not suitable for reproduction. In addition, it might be useful to look at some of the volumes already published or, especially if some atypical text is planned, to write to the Physics Editorial Department of Springer-Verlag direct. This avoids mistakes and time-consuming correspondence during the production period.
As a special service, we offer free of charge LaTeX and TeX macro packages to format the text according to Springer-Verlag's quality requirements. We strongly recommend authors to make use of this offer, as the result will be a book of considerably improved technical quality. The typescript will be reduced in size (75% of the original). Therefore, for example, any writing within figures should not be smaller than 2.5 mm.
Manuscripts not meeting the technical standard of the series will have to be returned for improvement.
For further information please contact Springer-Verlag, Physics Editorial Department II, Tiergartenstrasse 17, W-6900 Heidelberg, FRG.

A. Heck J. M. Perdang (Eds.)

Applying Fractals in Astronomy

Springer-Verlag
Berlin Heidelberg GmbH

Editors

André Heck
Observatoire Astronomique, 11 rue de l'Université
F-67000 Strasbourg, France

Jean M. Perdang
Institut d'Astrophysique de l'Université de Liège
Avenue de Cointe 5, B-4200 Cointe-Ougrée, Belgium

ISBN 978-3-662-13848-9 ISBN 978-3-540-47582-8 (eBook)
DOI 10.1007/978-3-540-47582-8

Originally published by Springer-Verlag Berlin Heidelberg New York in 1991
Softcover reprint of the hardcover 1st edition 1991

2153/3140-543210 - Printed on acid-free paper

Foreword

'Fractal geometry addresses itself to questions that many people have been asking themselves. It concerns an aspect of Nature that almost everybody had been conscious of, but could not address in a formal fashion.'

'Fractal geometry seems to be the proper language to describe the complexity of many very complicated shapes around us.'

(Mandelbrot, 1990a)

'I believe that fractals respond to a profound uneasiness in man.'

(Mandelbrot, 1990b)

The catchword *fractal*, ever since it was coined by Mandelbrot (1975) to refer to a class of abstract mathematical objects that were already known at the turn of the 19th century, has found an unprecedented resonance both inside and outside the scientific community. Fractal concepts, far more than the concepts of catastrophe theory introduced a few years earlier, are currently being applied not only in the physical sciences, but also in biology and medicine (Goldberger and West 1987).

In the mid-eighties, Kadanoff (1986) asked the question: *'Why all the fuss about fractals?'*. He offered a twofold answer: in the first place, it is *'because of the practical, technological importance of fractal objects'*. Indeed he emphasised the relevance of these structures for materials scientists and oil drilling engineers, in search of structures with novel properties, or models for the flow of oil through the soil. His second answer was: *'Because of the intellectual interest of fractals'*. And Kadanoff mentioned in this respect the invariance, or statistical invariance properties, of fractals under dilation and translation groups (to which we could add rotation groups as well).

It seems quite clear that if the idea of practical utility has been a decisive factor in promoting the diffusion of the new concepts, then fractals should be even more important for the observational sciences than for the experimental sciences. In fact we witness an ubiquity of fractal, or fractal-like, shapes in natural systems (Barnsley et al. 1988; Mandelbrot 1977; Peitgen and Richter 1986), which in man-controlled systems, including laboratory experiments, are largely artificially suppressed. Surprisingly, however, we notice that in the specific field of observational astronomy, fractals have been largely ignored – with the exception of the characterisation of the large-scale matter distribution in the universe, whose probable fractal nature was already stressed by Mandelbrot. The reason for this could be that, traditionally, astronomy relies on theoretically well-established concepts only; those concepts which are not yet part of a clearcut physical theory are typically rejected.

Kadanoff's concluding sentence, *'the physics of fractals is, in many ways, a subject waiting to be born'*, has remained true until recently. The answer to the central question: *'How are fractal structures generated in the framework of our conventional physics formulated in the language of differential or partial differential equations?'*, was not generally appreciated until perhaps the mid-eighties. It has been recognised only gradually that differential equations generically and inescapably produce fractals which are in fact responsible for the complex time-behaviour (chaos) exhibited by these systems (see also Devaney 1990).

Indeed, in the framework of physical applications, perhaps the now widest interest in fractal geometry stems from a need for a quantitative characterisation of the time-behaviour of nonlinear dynamical systems. Since the general realisation, in the seventies, of the irreductible character of nonlinearities in differential equations – the effect of small nonlinear corrections to linear equations cannot just be accounted for in the solutions by mere perturbation methods – fractal dimensions have been adopted systematically to capture the new complex geometrical pattern traced out by the attractors in the phase space of nonlinear dynamical systems. In the particular context of the analysis and of the modeling of *astronomical time series* (for instance brightness variations, velocity curves, and so on, of variable stars), a similar trend has developed only in the second half of the past decade, when the relevance of nonlinear behaviour was appreciated by a sufficient number of workers in this area (cf. Buchler et al. 1985). The topic of nonlinear astronomical time behaviour is now developing into an active field in which the concepts of fractal geometry play an important part.

As a by-product of recent work in chaos, it has now become widely recognised also that fractal space patterns are the standard outcome of generic 'flows' (cf. Ottino 1989). It is therefore natural to look for fractal structures in astronomical configurations which can be modelled by flow systems.

The idea of compiling this book came up during the *XIIth Strasbourg Astronomical Day* organized in 1990 on the theme *Fractals in Astronomy* (Heck 1990). It appeared then that fractals indeed had become a tool in their own right in the already quite vast astronomical methodology.

This volume is a compilation of review papers dealing with what we believe are the currently main astronomical and astrophysical areas in which fractals have been applied in a systematic way.

Perdang presents a general introduction to various aspects of fractal geometry, with particular emphasis being placed on the dynamical generation of fractal structures; the existing applications of fractals to astrophysical dynamical systems (oscillation models and analysis of time series) are critically analyzed. Goupil et al. concentrate on the specific fractal characterisation of strange attractors of stellar pulsations (especially in the case of White Dwarf model oscillations). Brandt et al. provide a critical overview of the attempts made by various authors to attach fractal dimensions to the solar granulation pattern; the physical mechanism motivating this characterization is turbulent convection, itself characterised by fractal geometry.

The next four chapters centre around the large-scale structure of the universe, which is dealt with from various points of view. Provenzale, relying like Brandt et al. on the idea of a turbulent mechanism, this time operating on a cosmic scale, adapts the β-model of fluid dynamic turbulence to obtain a geometric representation for the observed galaxy distribution. Calzetti and Giavalisco study the scaling properties of the angular and spatial two-point correlation function of galaxies.

Martínez, besides discussing a few empirical geometric models for the representation of the observed galaxy distribution, offers a detailed review of the currently applied methods of fractal analysis of the observational (point) data (including in particular the minimum spanning tree method). Rouet et al. develop an attempt at a theoretical interpretation of the observed hierarchical structure; it demonstrates that a 1D gravitational gas subject to the Hubble expansion undergoes a gravitational clustering leading to a spatially fractal distribution.

Nottale analyses the astrophysical consequences of a recent suggestion of substituting a fractal space-time manifold of general relativity.

Finally Heck reviews briefly the present situation in the field of astronomical databases and archives, with a few thoughts on shortcomings, future trends and potential methodological applications.

The goal of this volume would be fulfilled if it helped to stimulate the interest in applying fractal techniques in new areas of astronomy.

Acknowledgements

It is a real pleasure to thank all contributors for their prompt delivery of manuscripts which, together with the efficiency of the publisher, has made it possible for us to produce a volume with the timeliest and most up to date information available. Mrs. M. Hamm assisted us in problematic bibliographical identifications.

References

Barnsley, M.F., Devaney, R.L., Mandelbrot, B.B., Peitgen, H.O., Saupe, D., Voss, R.F. (1988): *Fractal Images* (Springer-Verlag, New York).

Buchler, J.R., Perdang, J., Spiegel, E.A. (Eds.) (1985): *Chaos in Astrophysics*, NATO ASI C **161** (D. Reidel Publ. Co., Dordrecht).

Devaney, R.L. (1990): *Chaos, Fractals, and Dynamics* (Addison-Wesley, New York).

Frisch, U. (1977): in *Problems of Stellar Convection*, eds. E.A. Spiegel, J.P. Zahn (Springer-Verlag, Berlin), p. 325.

Goldberger, A.L., West, B.J. (1987): *Amer. Scientist* **75**, 354.

Heck, A. (Ed.) (1990): *Fractals in Astronomy, Vistas Astron.* **33**, 245-424.

Khadanoff, L.P. (1986): *Physics Today* **39**, 6.

Mandelbrot, B.B. (1975): *Les objets fractals: forme, hasard et dimension* (Flammarion, Paris).

Mandelbrot, B.B. (1977): *The Fractal Geometry of Nature* (Freeman, New York).

Manbelbrot, B.B. (1990a): in *Fractals: an Animated Discussion*, Eds. H.O. Peitgen, H. Jürgens, D. Saupe, C. Zahlten (Spektrum der Wissenschaft, Heidelberg), videotape.

Mandelbrot, B.B. (1990b): ***Newsweek*** **26 March 1990, 56.**
Ottino, J.M.(1989): ***The Kinematics of Mixing: Streching, Chaos, and Transport*** **(Cambridge Univ. Press, Cambridge).**
Peitgen, H.O., Richter, P.H. (1986): ***The beauty of Fractals*** **(Springer-Verlag, Berlin).**

Contents

Geometry and Dynamics of Fractal Sets

Jean M. Perdang

Institute of Astronomy, Cambridge CB3 OHA, United Kingdom
and
Institut d'Astrophysique, B-4000 Liège, Belgium [1]

Abstract: The most relevant tools of fractal geometry are introduced, and several classes of mathematical frameworks which generate fractal configurations dynamically (in space-time as well as in parameter spaces) are examined.

0. Foreword

With the exception of the large-scale structure of the universe – an area in which the concept of a hierarchical organisation was around long before Mandelbrot's fractal geometry was created (cf. the observational investigations by de Vaucouleurs 1970, 1971, 1974, and the earlier theoretical speculations by Charlier 1922), and which, more recently, has experienced an enhanced activity through the introduction of fractal concepts (Davies and Peebles 1983, Bahcall and Soneira 1983, Jones et al. 1988, Bahcall 1988, Rood 1988, Peebles 1980, 1989, Coleman et al. 1988, Yoshioka and Ykeuchi 1989, Martinez and Jones 1990, Newman and Wassermann 1990, Pellegrini et al. 1990, Pierre 1990, Slezak et al. 1990) – references to fractal tools have been sparse among the astronomical community. It is the goal of this paper to emphasise the relevance of fractal ideas in other areas of astronomy where complicated spatial and temporal patterns are encountered which can be characterised by fractal dimensions. With a quantitative 'measure' of texture and structure at our disposal, it is often possible to step back from the morphology to the dynamics responsible for generating the observed patterns.

There exist at the moment various excellent introductory as well as more specialised texts on fractal geometry oriented towards applications to various fields of physics (Liu 1986; Pietronero and Tossati 1986; Feder 1989; Avnir 1989). Astrophysical applications are reviewed in Perdang (1990b). The mathematics of fractals is discussed in Falconer (1985) and Barnsley (1988). The topic of fractal image generation–of more than just aesthetic interest in astronomy–is reviewed in some detail in Peitgen and Saupe (1988).

---

[1] Permanent Address.

The fractal models we shall examine here are *dynamical models*, describing the evolution of physical processes in time, and thereby providing a causal interpretation of the origin of the fractal nature of the observed structures. Geometric algorithms and functions producing fractals, also referred to as 'fractal models' by Mandelbrot (1975, 1977, 1982), are conceptual frameworks useful for disclosing the interrelations of different properties of observed fractal-like physical systems. As we shall make it clear below, geometric reference models of any preassigned fractal dimension (or collection of dimensions) can be set up in a straightforward way. An elementary instance of this class of applications is the Sierpinski gasket in 3-dimensional space proposed as a model for the large-scale distribution of matter in the universe (cf. Calzetti et al. 1987, 1988). A less trivial example is Voss's fractal simulating a planetary surface. It is based on a geometric algorithm which can be summarised as follows: (0) start with a sphere of centre O and radius R (the planet's reference sphere); (1) select at random a great circle on the sphere; (2) in each hemisphere, labelled 0 and 1, determined by the great circle, let the altitude h execute one step of a (generalised) Brownian motion, so that the new radii of the two hemispheres become $R_0 = R + h_0$ and $R_1 = R + h_1$ respectively; this completes the first cycle of operations. Iterate steps (1) and (2) a large number of times; at iteration n the addition of an nth great circle determines $\leq 2^n$ spherical sections, which can therefore be labelled $d_1 d_2 \ldots d_{n-1} d_n$, $d_i = 0$ or 1 (if the nth circle does not cut section $d_1 d_2 \ldots d_{n-1}$, then set $d_n = 0$); the radius of each section is then given by $R_{d_1 d_2 \ldots d_{n-1} d_n} = R_{d_1 d_2 \ldots d_{n-1}} + h_{d_1 d_2 \ldots d_{n-1} d_n}$, i.e.

$$R_{d_1 d_2 \ldots d_{n-1} d_n} = R + h_{d_1} + h_{d_1 d_2} + \ldots + h_{d_1 d_2 \ldots d_{n-1}} + h_{d_1 d_2 \ldots d_{n-1} d_n} .$$

Illustrations of this model can be found in Voss (1988). Geometric models, interesting and useful as they may be as devices of image, and more generally, information compression, are not directly intended to address the question of the physics responsible for the fractal character. However, they may sometimes suggest a conceivable dynamical mechanism. Thus Voss's algorithm is indeed turned into a genuine dynamical model by merely re-interpreting the iterative cycles as 'geological timesteps'; the irregular planetary relief arises from an initially perfectly spherical surface which suffers 'quakes' at cycles $t_1, t_2, \ldots$; the cumulative process of a large number of quakes generates the fractal relief. We shall discuss how the elementary geometric construction of fractals can be recast into an algebraic formalism which in turn represents a dynamical system. Provided that the latter can be given a physical interpretation, the dynamical system supplies a physically conceivable evolutionary mechanism for the observed structure.

We shall not discuss the existing methods for describing physical processes over geometrical fractals by means of conventional partial differential equations. So far only a few attempts in this direction have been made (cf. O'Shaughnessy and Procaccia 1985 for diffusion; Perdang 1990c for gravitational fluid dynamics). As a rule, authors dealing with the physical behaviour over fractal sets resort to discrete schemes, in particular to *cellular automata* (cf. the discussion of localised oscillation modes in fractals; Courtens et al. 1989; Sapoval 1989). For a fractal approach to quantum cosmology carrying further Wheeler's (1962) idea of space-time 'foam' we refer to Nottale (1989; cf. also this volume).

1. From Pseudo-Dimensions to Fractal Dimensions

1.1. Pseudo-Dimensions

Let X be a global physical property referring to an object whose geometrical form is an n-dimensional set M embedded in a Euclidean N-dimensional space R^N, and suppose that under assumption A the property can be expressed as a formally continuous function H of the (integer) topological dimension n

$$X = H(n). \tag{1.1}$$

Suppose further that under a different assumption A', and for a fixed topological dimension n_o of the geometrical shape of the object, property X can be re-expressed in the same functional form (1.1), with n a formal, generally non-integer coefficient, defined by

$$n = f(n_o; A') \quad , \quad n \neq n_o \quad \text{if} \quad A' \neq A, \tag{1.1.a}$$

(f, continuous function of the parametrisation of the assumption A'). Then we refer to the parameter n as the X ***pseudo-dimension*** of the object of topological dimension n_o, under assumption A'. We shall denote an X pseudo-dimension by the symbol P_X. We may formally view the pseudo-dimension P_X as an index of the *geometry* of a shape A regarded as embedded in a Euclidean space of high enough dimension. Notice that in contrast to the original figure M which is a standard, typically smooth shape (an n-dimensional manifold), the formal geometrical set A may have an unconventional non-smooth structure. A few illustrative examples clarify this definition and indicate its potential interest.

The criterion for dynamical stability of a self-gravitating equilibrium figure

$$\gamma > \gamma_{crit}, \tag{1.2}$$

involves the adiabatic coefficient of the matter of the body, $\gamma = [\partial \ln P / \partial \ln \rho]_S$ where all symbols have their usual meanings, and a critical coefficient γ_{crit}, which, for a nonrotating n-dimensional spherically symmetric configuration embedded in an n-dimensional space (assumption A), obeys

$$\gamma_{crit} = 2\,(n-1)/n. \tag{1.3}$$

This formula indicates that for a self-gravitating configuration the degree of dynamical stability decreases with increasing dimension of the body: lowering the dimension means indeed reducing the number of degrees of freedom; the manifold of allowed perturbations under which the stability is to be tested thereby becomes a subset of the original manifold of perturbations.

For a homogeneous compressible 3-dimensional configuration rotating uniformly with angular velocity Ω (assumption A') one finds, from the approximate frequency formula for pseudo-radial oscillations (cf. Eq. 82.13 in Ledoux and Walraven 1958), that the critical stability coefficient is given by

$$\gamma_{crit} = 4/3 - (\Omega/\Omega_o)^2 \quad , \quad \Omega_o^2 = 6\pi G\rho, \tag{1.3.a}$$

to lowest order in the rotation (ρ, uniform density of the model). Identifying the right-hand sides of (1.3) and (1.3.a) we find a dynamical stability pseudo-dimension, P_{dyn}, for the rotating star

$$n = 3 - (9/2)(\Omega/\Omega_o)^2 \equiv P_{dyn} < 3. \tag{1.3.b}$$

The effect of a uniform rotation on the dynamical stability is thus equivalent to reducing the original topological dimension 3 of the star to a formal value (< 3). Intuitively we interpret this effect by observing that rotation, singling out a preferred plane of motion, renders the star 'more 2-dimensional'. By the same token we explain the stabilizing action of rotation. We also see that this result cannot be a peculiar feature of the homogeneous model (cf. Ledoux and Walraven 1958, Section 82). The actual adiabatic coefficient of a perfect classical gas can be written

$$\gamma = (f + 2)/f, \tag{1.4}$$

where f is the number of degrees of freedom of one particle, i.e. the *dimension of the configuration space* of the *classical* particle. In the *quantum* case we interpret f as an average number of excited degrees of freedom per particle, or a pseudo-dimension of a fictitious configuration space of a particle, $f \equiv P_{con}$. The dynamical stability condition of the rotating star expressed in both pseudo-dimensions takes the symmetric form

$$P_{con} < 2\,P_{dyn}/(P_{dyn} - 2) \quad \text{or} \quad P_{dyn} < 2\,P_{con}/(P_{con} - 2). \tag{1.5}$$

At fixed dimension of the star, P_{dyn}, [dimension of the gas, P_{con}], the configuration acquires a higher degree of stability as the dimension of the gas, P_{con}, [dimension of the star, P_{dyn}], is lowered, since we restrict the manifold of allowed perturbations.

As a second illustration take the observational mass-radius relation for the main sequence stars

$$\log(R/R_o) = A + B\log(M/M_o), \tag{1.6}$$

(A and $B = 1/d$, dimensionless constants). For the upper main sequence (O8-G4 stars), $B \approx 3/4$ (Parenago and Massewitsch 1951). On the other hand, if the thermonuclear energy generation $\epsilon(\rho, T)$ and the opacity $\kappa(\rho, T)$ are approximated by power laws in the temperature and density (assumption A')

$$\epsilon(\rho, T) = \epsilon_o\rho^\alpha T^\beta \quad ; \quad \kappa(\rho, T) = \kappa_o\rho^a T^b, \tag{1.7}$$

then it follows from the hydrostatic and thermal equilibrium equations that

$$M/M_o = K(R/R_o)^d, \tag{1.8}$$

with d given by

$$d = [3(a+\alpha) + (b+\beta)]/[(a+\alpha) + (b+\beta) - 2] \tag{1.8.a}$$

(cf. Hayashi et al. 1962). For the CNO cycle ($15 < \beta < 20$) d lies in the ranges 1.52 (low β) to 1.36 (high β) for the Kramers opacity, 1.49 to 1.34 for Schwarzschild's approximation, or 1.28 to 1.21 for electron scattering (cf. also Sears and Brownlee 1965 for various estimates). The mass $M(R)$ contained in an n-dimensional ball of radius R and of uniform n-dimensional density ρ_n (assumption A) obeys

$$M(R) = V(n)\rho_n R^n \ , \ V(n) = \frac{\pi^{\frac{n}{2}}}{\Gamma(1+n/2)} \tag{1.9}$$

($V(n)$, the n-dimensional 'volume' of the n-ball of unit radius). Expression (1.9) remains algebraically meaningful for arbitrary real values of n, so that the exponent d of (1.8) is a ***mass pseudo-dimension***, P_M. The pseudo-dimension of about 4/3 for the upper main sequence stars is less than the stars' topological dimension. This reflects the following physics. In the absence of nuclear energy generation a (classical) self-gravitating gas evolves towards a point configuration which is the eventual (classical) equilibrium state of the star; in this genuine equilibrium state, the star's dimension is therefore 0. The (proto)star starts out as an essentially homogeneous configuration, of mass pseudo-dimension 3. Stellar evolution is a gravitational collapse, from an initial homogeneous state towards a final singular mass point; it consists in a progressive, irreversible lowering of the mass pseudo-dimension (which we may surmise to occur monotonically with age). Intermediate stationary stages of the star, such as the main sequence phase, in which the gravitational collapse is halted by nuclear energy generation, should therefore be characterised by successively lower dimensions determined by the parameters of the energy generation and transport. This is precisely what the mass pseudo-dimension (1.8.a) expresses.

As a third example consider the diffusion equation in an n-dimensional homogeneous Euclidean space (assumption A). With an initial Dirac-distribution concentrated at the origin at ($t = 0$), the probability of return to the origin at time t for the diffusing matter (particle carrying out a random walk) is given by

$$P(t) = Kt^{-\frac{n}{2}}. \tag{1.10}$$

Numerical random walk experiments on physically homogeneous but geometrically irregularly shaped sets $\mathsf{A} \subset \mathsf{R}^{n_o}$ (assumption A') show that the probability of first return to the origin obeys a power law of form (1.10), with n depending on the morphology of the set A carrying the Brownian motion. Hence this exponent plays the part of a ***diffusion pseudo-dimension*** P_{diff} characterising the physical and geometrical properties of the set A. The latter dimension has been introduced by Alexander and Orbach (1982; cf. also Alexander et al. 1986) who refer to it as the ***fracton dimension***, or ***spectral dimension***. In fact, the wave equation of a homogeneous n-dimensional medium of arbitrary (regular) shape embedded in R^n admits of the asymptotic frequency distribution (Weyl 1911; Kac 1966, Baltes and Hilf 1978)

$$N(\omega) = K'\omega^n, \tag{1.11}$$

$N(\omega)$ standing for the number of modes of frequency $< \omega$. The standing waves in a physically homogeneous set A of irregular shape embedded in R^{n_o} obey Eq. (1.11) with n depending again on the geometry of the set A, (***spectral pseudo-dimension*** P_{sp}). From the formal equivalence between the wave equation and the diffusion equation one infers that the two pseudo-dimensions P_{diff} and P_{sp} of a set A should be identical (under conditions of physical homogeneity of the set). The number of acoustic frequencies of a star are found to obey a relation of type (1.11); in the case of a polytrope of index 3 the spectral pseudo-dimension is $P_{sp} \approx 4.7$ (Perdang 1982). Here the assumption of physical homogeneity is not obeyed: the sound speed in polytropic models vanishes at the surface of the star. In a star, therefore, the existence of a spectral pseudo-dimension differing from the topological dimension 3, reflects the inhomogeneities in the physics (surface singularity of the inverse sound speed) over a smooth standard geometrical configuration, rather than a shape of irregular geometry carrying homogeneous physics (as analysed in Alexander and Orbach 1982). We can anticipate, therefore, that a star's diffusion pseudo-dimension does not, in general, coincide with its spectral pseudo-dimension; the diffusion dimension is related to an inhomogeneity in the diffusion coefficient whose analytical behaviour differs, in general, from the behaviour of the sound speed. Under the conditions analysed by Alexander and Orbach (1982) (homogeneous physics over an irregular geometrical shape), Rammal and Toulouse (1983) show that $P_{spec} \leq P_M$. In the case of inhomogeneous physics over a regular configuration, the inhomogeneities in the properties defining the spectral distribution, and those in the density, are not correlated, so that there is no reason for this inequality to hold (mass pseudo-dimension of a real star expected to obey $0 \leq P_M \leq 3$; spectral dimension ≥ 3 for inverse sound speed with a surface singularity).

In these examples the pseudo-dimensions can be traced to a common mathematical origin. In the first illustration, the critical adiabatic coefficient is obtained when the adiabatic equations of motion are invariant under a one-parameter Lie group. In the second illustration the mass-radius power law is likewise a consequence of a Lie-group invariance of the equations of hydrostatic and thermal equilibrium. Pseudo-dimensions manifestly always arise in problems governed by differential or partial differential equations obeying an invariance under a one-parameter scaling group. If the physical functions vary monotonically over the relevant range of the physical variables then an ***approximate*** scaling invariance of the differential or partial differential equations is bound to hold over a restricted range of the physical conditions (in Barenblatt's 1979 terminology, it is associated with intermediate asymptotics). Since we are typically interested over restricted ranges of physical variables (say the upper main sequence, etc) the occurrence of pseudo-dimensions in observational problems is not surprising.

A further general connection between the pseudo-dimension – and hence the formal geometrical set of dimension P_X – and the physics in the original space is made clear by the specific mass-radius relation. By Eq. (1.9) in which n is set equal to the topological dimension of the star, define an associated 3-dimensional density, $\rho_3(R)$ (average density of a main sequence star of radius R):

$$\rho_3(R) = [V(4/3)/V(3)]\rho_{4/3}R^{-\frac{5}{3}}. \tag{1.12}$$

The coefficient ρ_d ($= \rho_{4/3}$) is independent of R, so that it represents a ***homogeneous density*** over a formal ball-like geometrical configuration. For noninteger values of d such a set is referred to as a *fractal ball* of dimension d. We can visualise this figure as a sponge embedded in R^3, with an approximate spherical symmetry around a centre O, and such that the mass $M(R)$ contained in the intersection of this set by a 3-dimensional ball of radius R measured in the embedding space obeys the power law (1.9). Note that the geometrical nature of the spongy structure is not uniquely specified by this latter condition. The example indicates that a conventional ball in the 3-dimensional Euclidean configuration space carrying an inhomogeneous matter distribution (described by a space dependent conventional density) is formally equivalent to a d-dimensional fractal ball (regarded as embedded in R^N, N an integer) and whose matter distribution is homogeneous, provided that the original 3-dimensional density obeys a power law in the radius. As a result, the 3-dimensional density either vanishes, or is singular at the origin $R = 0$; under the first alternative the carrier of the homogeneous density has a dimension $d > 3$, under the second $d < 3$. A similar interpretation applies to formula (1.11), which is a 'mass' formula for an n-dimensional ball of radius ω, in a formal wave-number space.

1.2. Fractal Mass (Capacity) Dimension

A pseudo-dimension characterises a global physical magnitude X defined over a set of standard geometry (in general an n-dimensional manifold M^n); as a rule, property X is *inhomogeneously* distributed over the set. Our illustrations show that we can likewise view X as being defined over an abstract geometrical set $\mathsf{A} \subset \mathsf{R}^N$, over which X is then *uniformly* distributed. The pseudo-dimension appears as a parameter describing the geometrical structure of the abstract set. We here adapt this parameter to describe a geometrical shape of *arbitrary irregularity* (non-differentiable, etc); a complementary approach based on a discrete representation of the geometrical sets is presented in Perdang (1990b).

Let A represent an arbitrary geometrical set in R^N (support of stellar matter in the 3-dimensional space; attractor in the phase space of a model for the nonlinear oscillations of a star; boundary in the instability strip of the HR diagram separating regular oscillations from chaotic ones, etc). Imagine a unit mass uniformly spread over this set. Let $x \in \mathsf{A}$ be a reference point in the set. Let $\mathsf{B}^N(R;x)$ be the N-dimensional ball of centre x and radius R. Denote by $M(R;x)$ the mass contained in $\mathsf{A} \cap \mathsf{B}^N(R;x)$. If for R small enough a mass pseudo-dimension P_M exists, i.e. $M(R;x)$ obeys a power-law (Eq. 1.8)

$$P_M = \lim_{R\to 0} \ln M(R;x)/\ln R \equiv F_M(x), \tag{1.13}$$

then we refer to the local mass pseudo-dimension as the *fractal mass (capacity) dimension* of the set A at point x. If the set A is a standard smooth n-manifold, M^n, then the fractal mass dimensions $F_M(x)$ coincide with the topological dimension

n. If the fractal mass dimensions $F_M(x)$ do not coincide with the topological dimension then A is called a *fractal set.* If we have

$$\forall x \in A \quad : \quad F_M(x) = F_M \tag{1.14}$$

regardless of the reference point x, then the parameter F_M is the *fractal mass (capacity) dimension* of the set A.

A standard n-dimensional manifold $\mathsf{M}^n \subset \mathsf{R}^N$ has locally the structure of an n-dimensional Euclidean space, or equivalently the neighbourhood of x in M^n, $\mathsf{M}^n \cap \mathsf{B}^N(R;x) = \mathsf{B}^n(R;x)$ for R small enough, is an n-ball; instead of the N-ball we may use an N-dimensional hypercube of centre x and side R, $\mathsf{I}^N(R;x) = (x_1 - R/2, x_1 + R/2) \times \ldots \times (x_N - R/2, x_N + R/2)$; we then have $\mathsf{M}^n \cap \mathsf{I}^N(R;x) = \mathsf{I}^n(R;x)$ for R small enough. It is this locally Euclidean behaviour which enables us to define the n-dimensional measure of a standard manifold M^n, $\mathcal{M}_n(\mathsf{M}^n)$ – length, area, volume, ... if $n = 1, 2, 3, \ldots$ respectively – as the sum of the measures of a large enough number of n-hypercubes [n-balls] covering the manifold (the n-measure of an n-hypercube $\mathsf{I}^n(R;x)$ is R^n; the n-measure of an n-ball is given by Eq. 1.8). In a similar way, according to our definition, the point $x \in \mathsf{A} \subset \mathsf{R}^N$ has a neighbourhood which is a fractal ball of mass pseudo-dimension $F_M(x)$: $\mathsf{A} \cap \mathsf{B}^N(R;x) = \mathsf{B}^{F_M(x)}$; we could equivalently use $\mathsf{A} \cap \mathsf{I}^N(R;x) = \mathsf{I}^{F_M(x)}$, where the set $\mathsf{I}^N(R;x)$ denotes a formal generalisation of a hypercube (fractal hypercube) whose F_M-measure is R^{F_M}. If the set A has a global fractal mass dimension F_M (Eq. 1.14), then we can attach an F_M-dimensional measure to this set, $\mathcal{M}_{F_M}(\mathsf{A})$, by the natural generalisation of the procedure used in conventional manifolds. This measure, referred to as the *exterior measure* [2], has been introduced by Hausdorff (1919; cf. Falconer 1985). We leave out the subscript F_M to the symbol $\mathcal{M}$ if there is no ambiguity. Notice that it has the physical 'dimension' $(\text{length})^{F_M}$.

Hausdorff's dimension F_M of a set A parallels the topological dimension also in the following respect. Denote by

$$L^n = \{x \,|\, x = \lambda_o x_o + \lambda_1 x_1 + \ldots + \lambda_n x_n, \; x_i \in R^N, \text{ linearly independent},$$
$$\lambda_i \in R^1, \quad i = 0, 1, \ldots, n\}$$

a linear n-dimensional subspace of the Euclidean embedding space R^N. The *join* of two linear subspaces L^n and L^m of defining points $x_o, x_1, \ldots, x_n$ and $y_o, y_1, \ldots, y_m$ respectively is the set of points linearly dependent on the latter two sets of points. The dimension of the intersection (d_I) of L^n and L^m in R^N is given by

$$d_I = n + m - d_J, \tag{1.15}$$

where d_J is the dimension of the join of the two spaces (cf. Todd 1947). If the defining points are in general position [3] and if $n + m + 1 \leq N$, then all $n + m + 2$ points are linearly independent. Hence $d_J = min\{n + m + 1, N\}$. Therefore

[2] The dimension F_M is sometimes called *exterior dimension* (cf. Zeldovich et al. 1990).

[3] By this we mean here that the position of the points is such that under an infinitesimal perturbation of the coordinates the linear dependence or independence of the points is not changed.

$$d_I = -1, \quad \text{if } n+m+1 \leq N; \; d_I = n-(N-m), \quad \text{if intersection } \neq \emptyset \tag{1.15.a}$$

(mass pseudo-dimension of the intersection) with the standard convention that the empty set is given the topological dimension −1 (cf. Favard 1950); the quantity between brackets is the codimension c of the subspace L^m (dimension of the orthogonal complement, L^{N-m}, to L^m). Since an n-dimensional smooth manifold M^n is locally a linear space L^n, (1.15.a) holds for arbitrary smooth manifolds (assuming again that the manifolds are in general position). Substitute next a fractal set A in R^N for the smooth manifold L^n. By selecting a point x in the intersection of A with a manifold L^m, and considering the ball $\mathsf{B}^N(R;x)$ one can convince oneself that this intersection has a fractal mass dimension given by relation (1.15.a)

$$F'_M = F_M - (N-m), \tag{1.16}$$

where F_M and F'_M are the fractal dimensions of A and of the intersection respectively, provided again that the intersecting sets are in general position (see Mandelbrot 1982, Section 14 for comments on exceptions to rule 1.16). [4]

Finally, it is immediately seen that the Cartesian product set $\mathsf{A} \times \mathsf{M}^m$, $\mathsf{A} \subset \mathsf{L}^{n'}$ and $\mathsf{M}^m \subset \mathsf{L}^{N-n'}$, has a fractal mass dimension

$$F''_M = F_M + m, \tag{1.17}$$

which extends the addition formula of topological dimensions of Cartesian products (for the direct product of two fractal sets see Falconer 1985 chapter 5).

If the fractal mass dimension is both independent of the position x in the set, and of the scale R, then the mass of A intercepted by an N-dimensional ball of radius R, $M(R)$, obeys the mass-radius power law with exponent $F_M(= P_M)$ for any R (over some finite range), or the measure $\mathcal{M}_{F_M}$ of these intercepts scales with R as R^{F_M}. Let A' be the transformed set A under the similarity $\{\mathsf{O}, \lambda\}$, in the R^N embedding space of A, (with every point $x \in \mathsf{A}$, of coordinates $(x_1, x_2, \ldots, x_N)$ in a Cartesian coordinate frame with origin O, we associate a point x' of A', of coordinates $(x'_1, x'_2, \ldots, x'_N) = \lambda(x_1, x_2, \ldots, x_N)$, λ a positive real). It will be convenient to use the notation $\lambda\,\mathsf{A}$ for any set congruent to the transformed set A'. Denote by $(\mathsf{A})_R$ the the intersection of A with $\mathsf{B}^N(R;x)$. Then the F_M-dimensional measure obeys

$$\mathcal{M}(\lambda(A)_R) = \lambda^{F_s}\mathcal{M}((A)_R), \tag{1.18}$$

where we have written F_s for the fractal mass dimension F_M. This particular dimension F_s is referred to as the ***similarity dimension***. If the geometry of the set A is such as to obey

[4] Relation (1.16) may be regarded as holding also for an empty intersection, if one interprets the modulus of the next highest integer of a negative dimension (topological or fractal) as the minimum number of independent parameters one needs to change to produce a non-empty intersection (Perdang 1988): for a straight line and a single point in general position in 3-space, relation (1.16) gives -2; by changing two of the 3 coordinates of the point, the latter is made to coincide with the line. See Cates and Deutsch (1987), Fourcade et al. (1987), Prasad et al. (1988) and Mandelbrot (1989) for other comments on negative dimensions.

$$\lambda(A)_R = (A)_{\lambda R}, \tag{1.19}$$

for *some* scaling factor λ, then the set is *self-similar*, or *hierarchical.* From (1.18) we have for a hierarchical set

$$F_s = \ln \mu / \ln \lambda, \tag{1.20}$$

with

$$\mu = \mathcal{M}((A)_{\lambda R})/\mathcal{M}((A)_R). \tag{1.20.a}$$

This relation is used for computing the fractal mass dimension of simple hierarchical sets for which the scaling ratios of the lengths, λ, and of the 'masses', μ, can be directly read off from the geometrical structure (construction algorithm).

Suppose that the set A is given (sampled) by a cloud of points,

$$A \approx \{x_1, x_2, \ldots, x_T\} \tag{1.21}$$

(x_j, $j = 1, 2, \ldots, T$ defined in R^N, with T large enough). Theoretical sets defined by an iterative algorithm are approached numerically in a finite number of iterative steps; they are naturally approximated by a finite number of points, or lines, or elementary areas (triangles), etc; by using the endpoints of the lines, the vertices of the triangles, etc we obtain a sampling of form (1.21). Observational sets are typically recorded in the form of a digitised picture or discrete time series (finite point sets). If the points x_i cover the set *uniformly*, then we have $M(x; R) \approx \nu(x; R)/T$, with $\nu(x; R)$ the number of sampling points in an N-ball of radius R centred at a reference sampling point x. Equivalently, $\nu(x; R)$ may represent the number of sampling points in the N-hypercube of side R and centre x. From Eq. (1.13) it follows that for R small enough

$$\nu(x; R) = K R^{F_M}, \tag{1.22}$$

(where K and F_M may depend on the reference point x). The slope of the plot of $\ln \nu(x; R)$ against $\ln R$ extrapolated to small R values then gives the fractal mass dimension at the reference point x. Such a plot is practically meaningful only for radii $R > r$ such that the corresponding number of points $\nu(x; r)$ remains large enough (ideally > 100). On scales $R <$ average distance between sampling points, the method reproduces the true fractal mass dimension of a finite point set, namely 0. An extension of this procedure provides a direct estimate of the global fractal mass dimension. Partition the embedding space R^N into a regular network of N-dimensional hypercubic cells of side r,

$$I^N(r; O + (k_1 e_1 + k_2 e_2 + \ldots k_N e_N) r), \; k_1, k_2, \ldots, k_N \text{ integers},$$

$\{O; e_1, e_2, \ldots, e_N\}$ a Cartesian coordinate system of origin O and unit vectors along the coordinate axes $e_1, e_2, \ldots, e_N$. The cells covering the set A (i.e. $\mathsf{I}^N(r; x) \cap \mathsf{A} \neq \emptyset$), are then labelled $i = 1, 2, \ldots, Q(r)$, $Q(r)$ total number of cells needed at 'resolution' r. For a set of fractal mass dimension F_M and total mass normalised to 1

$$1 = \sum_{i=1}^{Q(r)} \nu(x_i; r) \approx Q(r) K r^{F_G}, \tag{1.23}$$

with $\nu(x_i; r)$ given by Eq. (1.22) under the assumption that the cells are uniformly covered. Eq. (1.23) may continue to hold if the assumption of uniform cell occupation is violated; therefore we denote the exponent by F_G. The number of occupied 'boxes', or grid cells, then scales with resolution, for r small enough, as

$$Q(r) = K'(1/r)^{F_G}, \tag{1.24}$$

F_G being the *fractal grid dimension* (or *box dimension*) [5] (negative slope of a log - log plot of the number of covering boxes, $Q(r)$, against resolution r;*Box Counting Algorithm*). The mass dimension of the power spectrum of a time series has been proposed as an indicator of the underlying dynamics (regular or chaotic oscillation; Perdang 1981, Blacher and Perdang 1981).

These methods apply to the calculation of the fractal mass dimension of an *attractor*, i.e. the geometrical set carrying an asymptotic solution $x(t)$ of a dynamical system (defined by difference equations, differential equations, or partial differential equations) in an N-dimensional Euclidean phase space. If the trajectory is computed at discrete times $t_i, i = 1, 2, \ldots, T$, $x(t_i) = x_i$, then the collection of points $\{x_1, x_2, \ldots, x_T\}$ supply the desired sampling (1.21) of the attractor, providing a full coverage of the latter if the trajectory is known over a long enough time interval. In the case of a low-dimensional phase space, a visual inspection of the plot of the projection of the orbit on the coordinate planes indicates whether this condition is fulfilled. *Uniformity* of the sampling over the set is an assumption that remains to be tested (cf. Mayer-Kress 1987 for a discussion of the practical difficulties encountered with the estimates of attractor dimensions; the number of sampling points needed for an evaluation of the dimension increases exponentially with the dimension to be estimated).

1.3. Correlation Dimension

If a set A given in the sampled form (1.21) is known to admit of a global fractal mass dimension F_M, then the latter can be obtained by a method devised by Grassberger and Procaccia (1983a, 1983b). Rewrite the 'correlation integral'

$$C(R) = (1/T^2) \sum_{i,j=1}^{T} \Theta(R - |x_i - x_j|), \tag{1.25}$$

($\Theta(x)$, step function, $\Theta(x) = 0$ for $x \leq 0$, and 1 for $x > 0$), as

$$TC(R) = (1/T) \sum_{j=1}^{T} \nu(x_j; R); \ \nu(x_j; R) = \sum_{i=1}^{T} \Theta(R - |x_i - x_j|), \tag{1.25.a}$$

[5] The term *box dimension* is sometimes reserved to a procedure of covering the set by a minimal number of N-dimensional boxes of side r; cf. Voss 1988.

(ν, number of points of the cloud inside the ball of radius R centred at the reference point x_j). Therefore, by Eqs. (1.22) and (1.14) we have, for R small enough

$$C(R) = BR^{F_C}, \tag{1.25.b}$$

with B independent of R. The exponent F_C is equal to the fractal mass dimension, $F_C = F_M$ if the sampling is uniform. Let T now tend to infinity in (1.25). If a scaling relation (1.25.b) holds, regardless of whether the sampling is uniform or not, then the exponent F_C is referred to as the *correlation dimension.* A popular graphical method for estimating F_C of an attractor consists in plotting the logarithmic derivative d ln $C(R)$ / d ln R against ln R. The occurrence of a plateau in this plot gives the value of the correlation dimension F_C, if it exists. Note that the correlation dimension is not an intrinsic parameter of the geometry of a set: it characterises a distribution of a sampling points over the set. Accordingly, an estimate of the correlation dimension of an m-manifold does not necessarily coincide with m; there is no guarantee for the correlation dimension to preserve the simple properties (1.16) and (1.17) under intersections and direct products.

1.4. Generalised Dimensions

The correlation dimension is one representative of a continuous class of dimensionalities attached to ***distributions of sampling points*** over geometrical sets A, known as *generalised dimensions* (Hentschel and Procaccia 1983, Grassberger 1985; see also Rasband 1990). To introduce the latter suppose again that the embedding space $\mathsf{R}^N \supset \mathsf{A}$ is covered by a regular network of cells of resolution r, as in the box counting algorithm. Denote by T_i and $P_i = T_i/T$ the total number and the relative number of sampling points in cell i respectively. The collection $\mathbf{P} = \{P_1, P_2, \ldots, P_{Q(r)}\}$ is a probability distribution over $Q(r)$ states. Following Jaynes (1957) the *information content* of the distribution is defined by

$$J(r;\mathbf{P}) = \mathbf{P} \bullet \log_2(\mathbf{P}Q(r)) = \log_2 Q(r) + \sum_{i=1}^{Q(r)} P_i \log_2 P_i. \tag{1.26}$$

The information content is thus minimum and vanishes if all states are equiprobable, and maximum if one state has probability 1, so that it obeys

$$0 \le J(r;\mathbf{P}) \le \log_2 Q(r) \tag{1.26.a}$$

The *Shannon information* $I(r;\mathbf{P})$ is the maximum information content minus the actual information content of the probability distribution

$$I(r;\mathbf{P}) = -\mathbf{P} \bullet \log_2 \mathbf{P} = -\sum_{i=1}^{Q(r)} P_i \log_2 P_i \tag{1.27}$$

and hence a measure of the *missing information* in the probability distribution. Since the latter behaves as a logarithm of the number of boxes $Q(r)$, we are naturally led to introduce an asymptotic scaling parameter for the Shannon information in the form

$$F_I = \lim_{r \to 0} I(r;\mathbf{P})/\log_2(1/r). \tag{1.27.a}$$

This parameter is the *information dimension* of the set A. Substitute next to the Shannon information the following expressions (*Renyi informations*)

$$I^{(q)}(r;\mathbf{P}) = \lim_{s \to q} 1/(1-s)\ \log_2 \sum_{i=1}^{Q(r)} P_i^{s} \tag{1.28}$$

(Renyi 1970), with $I^{(1)}(r;\mathbf{P}) = I(r;\mathbf{P})$ the standard Shannon information. Associated with the Renyi informations we have a continuum of information dimensions

$$F(q) = \lim_{r \to 0} I^{(q)}(r;\mathbf{P})/\log_2(1/r), \tag{1.28.a}$$

known as *generalised dimensions* or *Renyi dimensions.* Pawelzik and Schuster (1987) indicate that the latter are recovered from a correlation type integral (1.25)

$$C^{(q)}(R) = \{(1/T)\sum_{i=1}^{T}[(1/T)\sum_{j=1}^{T}\Theta(R-|x_i-x_j|)]^{q-1}\}^{1/(1-q)} \tag{1.29}$$

$(T \to \infty)$, which obeys the scaling relation, for R small enough

$$C^{(q)}(R) = K''R^{F(q)} \tag{1.29.a}$$

We have $F(1) = F_I$; $F(0) = F_G$ (grid dimension or fractal mass dimension); and $F(2) = F_C$. The derivative of (1.28.a) with respect to the order q is shown to be negative (cf. Schlögl 1980), so that the Renyi information is a non-increasing function of q (cf. Beck 1990 for a closer discussion of the inequalities obeyed by the Renyi dimensions). Therefore the generalised dimension is a decreasing function of q

$$F_G \geq F_I \geq F_C, \tag{1.29.b}$$

with the equality sign holding for a uniform sampling ($P_i = 1/Q(r)$) (cf. above).

We mention a final fractal dimension specifically attached with attractors of dynamical systems (Grassberger 1985, Rasband 1990). Let $L_i, i = 1,2,...,N$, denote the Lyapunov exponents of a trajectory $\mathbf{x}(t)$ of a dynamical system in the N-dimensional phase space. The latter measure the local convergence [divergence] of the departures $\delta\mathbf{x}(t)$ from the trajectory $\mathbf{x}(t)$ (the component δx_i along the eigenvector of L_i evolves as $\sim e^{L_i t}$), the occurrence of positive Lyapunov exponents characterising the 'sensitive dependence on initial conditions' of chaotic behaviour. Order these coefficients as follows

$$L_1 \geq L_2 \geq \ldots \geq L_u \geq \ldots \geq L_N,$$

with u the integer such that $L_1 + L_2 + \ldots + L_u \equiv S_u \geq 0$, and $S_u + L_{u+1} < 0$, and $u = 0$ if $L_1 < 0$. Then the parameter

$$u - S_u/L_{u+1} = F_L \tag{1.30}$$

has the following property: if the attractor is a manifold M^n (equilibrium point $n = 0$; limit cycle $n = 1$; torus $n = 2$, etc) then F_L is equal to the topological dimension n (for a smooth attractor the number of zero Lyapunov exponents is equal to the dimension). Since expression (1.30) remains meaningful in the presence of positive Lyapunov numbers (corresponding to 'strange' attractors), it can be regarded as defining a pseudo-dimension characterising the geometry of strange attractors. This parameter is known as the *Lyapunov* or *Kaplan-Yorke dimension.* Lyapunov dimensions have not been estimated in an astrophysical context. It has recently been argued that they are reliably obtained from observational time series (cf. Briggs 1990).

1.5. Multifractal Measures: Spectrum of Singularities

While a dispersion of the generalised dimensions indicates that the sampling of the set is not uniform, the knowledge of the spectrum of dimensions $F(q)$ does not directly provide an information on the statistics of the local fractal mass dimensions associated with the distribution of points. We introduce therefore another characterisation which refers directly to the statistics of the local fractal mass dimensions. This information is actually contained in the generalised dimensions, as will be shown in this section. Introduce a 'thermodynamic potential'

$$\tau(q) \equiv (q-1)F(q), \tag{1.31}$$

defining a curve in the $\{q, \tau\}$ plane. Since $F(q)$ is a decreasing function of q, this potential has a single extremum. Therefore, as in thermodynamics, we can select as a new independent variable the conjugate to q with respect to the potential,

$$\alpha = d/d\tau(q), \tag{1.31.a}$$

and consider the Legendre transform

$$\tau(q) - q\, d/dq\, \tau(q) \equiv -f(\alpha), \tag{1.32}$$

which is a function of the conjugate variable α alone. Expression (1.31) defines the same geometric curve by means of the local slope, $\alpha(q)$, and the point of intersection of the tangent with the τ axis, $f(\alpha)$. The function $f(\alpha)$, on the other hand, just like the Legendre transforms of the internal energy in thermodynamics, preserves the original information on the probability distribution, and has a single extremum in α: it has a parabolic shape of maximum given by the grid dimension (cf. 1.32)

$$\max \quad f(\alpha) = F(0). \tag{1.32.a}$$

The width of the parabola gives an indication of the scatter of the dimensions. The function $f(\alpha)$ is the *multifractal spectrum*, or *spectrum of singularities*, since it characterises the distribution of the points over the cells. In fact, approximate the occupation probability of cell i, at a given fine enough resolution r, in the form

$$P_i = A\, r^{\alpha_i}, \tag{1.33}$$

(A, normalisation constant). A large [small] exponent α_i thus characterises a low [high] concentration of points. The width of the interval of exponents $[\alpha_m, \alpha_M]$ is a measure of the lack of homogeneity of the distribution. We shall be interested in the statistical distribution of the exponents over the interval $[\alpha_m, \alpha_M]$, rather than in their individual values. To this end we represent $dn(\alpha)$, the number of exponents in $[\alpha, \alpha + d\alpha)$, by the scaling relation

$$dn(\alpha) \propto d\alpha\, r^{-f(\alpha)}, \tag{1.34}$$

in which $f(\alpha)$ specifies the distribution of the exponents i.e. the singularities. In terms of this distribution, the summation over the cells in the defining relation of the Renyi information (Eq. 1.28) becomes an integral over the distribution of the exponents α_i:

$$\sum_{i=1}^{Q(r)} P_i{}^q \rightarrow \int dn(\alpha)\, P(\alpha)^q \propto \int d\alpha\, r^{-f(\alpha)+q\alpha}. \tag{1.35}$$

The exponent of r is the 'potential' $\tau(q)$ (Eq. 1.32). Definition (1.28.a) implies that $\tau(q)$ relates to the generalised dimensions $F(q)$ by (1.31) (cf. Halsey et al. 1986).

From the generalised dimensions we thus obtain the spectrum of singularities $f(\alpha)$ (Eqs. 1.31,31.a,32), which is an *indicator of the concentration of the sampling points* over the subsets of A (distribution of mass pseudo-dimensions). Of course, a similar information is obtained by computing directly the local fractal mass dimensions $F_M(x)$ (Eq. 1.13) over a large number of points x. For a more detailed review of multifractal measures see Feder (1989). Arnéodo et al. (1988) have shown that the wavelet transform is an efficient tool for the calculation of $f(\alpha)$.

The knowledge of the spectrum of singularities, $f(\alpha)$, or alternatively the direct evaluation of the distribution of the local dimensions $F_M(x)$ is particularly relevant in the characterisation of attractors, for which the nature of the distribution of the sampling points is not known a priori. In the astrophysical context an attempt at estimating $f(\alpha)$ for the attractor of the X-ray variability of Her X-1 has been made by Atmanspacher et al. (1988).

1.6. Attractors

With chaotic dynamics as a potential framework for interpreting the complex observational time series of various astrophysical phenomena (light curves or velocity curves of Variable Stars, etc), the problem of reconstructing the attractor from a given time series has received considerable attention in the literature. The reconstruction from a time series $S(t), t = i\Delta t, i = 1, 2, ..., T + (N-1)s$, is based on two steps (cf. Casdagli 1989). (1) The actual phase space variables $\mathbf{x}$, and the (topological) dimension of the effective phase space of the phenomenon whose signal $S(t)(\equiv F(\mathbf{x}(t)))$ is observed, are generally unknown; therefore, a finite trial dimension N is chosen, and a fictitious N-dimensional Euclidean phase space vector is introduced by the method of finite delay

$$\mathbf{y}(t) = (S(t), S(t+\tau), S(t+2\tau), \ldots, S(t+(N-1)\tau)); \tag{1.36}$$

the delay $\tau = s\Delta t$ is arbitrary in principle; in practice it is chosen equal to 1/4 of a cycle of the signal. (2) If the observed signal corresponds to an asymptotic evolution, then $\mathbf{x}(t)$ evolves on the attractor A in the true phase space of the system; $\mathbf{y}(t)$ then generates an image A' of the attractor in the trial phase space. By Whitney's theorem any smooth manifold of dimension m, M^m, can be embedded in a Euclidean phase space R^N of dimension

$$N = 1 + 2m \tag{1.37}$$

in the sense that there exists a homeomorphism $H : \mathsf{M}^m \rightarrow \mathsf{M}^{m'} \subset \mathsf{R}^N$ (cf. Favard 1950). Since we expect that any set A of fractal mass dimension F_M can be carried by a smooth manifold M^m of dimension

$$m = 1 + \mathrm{int}\, F_M, \tag{1.37.a}$$

(int, integer part of F_M), *any* attractor of fractal dimension F_M is embeddable in a Euclidean space of dimension N given by (1.37,37.a) (cf. Takens 1981 for a rigorous proof). Whitney's modified theorem thus supplies the lowest dimensional R^N capable of carrying *any* set A of fractal dimension F_M. [6] In order for the transformation of the smooth manifold M^m, support of the attractor A in the true phase space (A carrier of the asymptotic phase points $\mathbf{x}(t)$), to the manifold M'^m supporting A' in the trial phase space (A' carrier of the asymptotic phase points $\mathbf{y}(t)$) to define a homeomorphism it is clear that (a) the trial phase trajectories must be non-intersecting; (b) the dimension N must be high enough so that the fractal dimension F_M of the true attractor A be compatible with Whitney's theorem. The estimate of the fractal dimension of the image A' in the trial phase space R^N must not exceed the value F_M given by (1.37,37.a).

So far virtually all authors who have analysed astronomical time series have applied the method of the correlation dimension F_C (1.25) of the trial attractor A'. The technique consists in plotting a family of $d/d\ln R \ln C(R)$ - $\ln R$ curves, for different values of the trial dimension N; if the family reveals a stable plateau (appearing for any N greater than some minimum dimension), then the latter is identified with the fractal dimension of the true attractor A. Such an identification is subject to several criticisms. In the first place, the fractal dimensions are *metric* properties, and therefore they are not topological invariants: if A and A' are homeomorphic, then their fractal dimensions are generally not equal. Secondly, as has been shown by numerical experiments by Cannizzo et al. (1990), even a stable plateau may be an artifact of the data: analysing the effect of a diffeomorphism in the trial phase space ($\mathbf{y}(t) \rightarrow \mathbf{z}(t)$ an elementary function of $\mathbf{y}$) these authors observe that the estimated correlation dimension of the new image of the attractor A'' may differ completely from the estimate of A' (they find a change from 1.6 to 3.6 when using exponentials of the previous variables); the difference results from a change in the probability distribution of the sampling points over A' and A''

---

[6] It may happen that attractors of simple enough geometry can be embedded in lower dimensional phase spaces.

which is strongly affected by a nonlinear change of variables. Thirdly, the correlation dimension is more sensitive to the distribution of the sampling points than for instance the mass dimension. These observations signal that estimates of the dimension of the true attractor drawn from time series via the single determination of F_C of a trial attractor remain problematic. Additional tests, such as the non-intersection test (a), should be carried out. Moreover, the singularity spectrum $f(\alpha)$ should be determined, or a direct estimate of the statistical distribution of the local mass dimensions should be made.

In the case of stellar pulsations, Kovacs and Buchler (1988) have estimated F_C from the time series of the stellar radius $R(t)$ of their full hydrodynamic models of Population II Cepheids; for all models tested the dimension lies in the range 1.9 to 2.5. One may surmise that in the case of theoretical models the uncertainties of the method are minimised (the radius is a good phase space variable sampling the attractor approximately uniformly, so that F_C remains close to F_M). Moreover these estimates agree with the dimensions of the attractors of elementary one-zone models developed by various authors (cf. Perdang 1991). In the observational context, a recent analysis of the visual lightcurve of R Scuti by Kollath (1990) has led to a correlation dimension between 4 and 5 (significantly higher than the theoretical model attractor dimensions). Besides the difficulties already mentioned, estimates of such a high dimension are intrinsically unreliable as a consequence of the limited number of data points. A search for a dimension of the attractor of white dwarf oscillations was made by Auvergne and Baglin (1986), and for the variability of Mira, R Leonis and V Bootis by Cannizzo et al. (1990).

In the case of observational signals resulting from astrophysical mechanisms which are only partially understood and which involve more complex physics than the dynamical oscillations of classical stellar variables, several authors have attempted to estimate attractor dimensions. Such dimensions provide constraints on the dynamics responsible for the signal (effective number of degrees of freedom). The X-ray luminosity of Her X-1 has been analysed by Atmanspacher and coworkers (Voges et al. 1987, 1988; Atmanspacher et al. 1988) who obtain $F_C \approx 2.3$ (cf. the objections by Norris and Matilsky 1989, and by Cannizzo et al. 1990); the variability of Cygnus X-1 is analysed in Lochner et al. (1989). The correlation integral for the outbursts of the Cataclysmic Variable SS Cygni reveals no stable dimension (Cannizzo and Goodings 1988).

2. Construction of Hierarchical Sets. Associated Fractal Dimensions.

We analyse here the geometric construction of a versatile class of model fractals, namely the *deterministic* and *statistical hierarchical sets* by means of an algebraic construction based on an iterative transformation which prepares the dynamical models to be discussed below. For a pure geometric construction see Mandelbrot

(1977, 1982; cf. also Perdang 1988). A minor modification of our iterative transformation changes the latter into Barnsley's iterative function system (IFS). [7]

2.1. Deterministic and Statistical Hierarchical Sets

Let B be a basic set (m-simplex B^m of the embedding space R^N , $m \leq N$, specified by the $m+1$ linearly independent vertices $(\mathbf{y}_1, \mathbf{y}_2, \ldots, \mathbf{y}_{m+1})$ in the unit closed hypercube centred at the origin, $\mathsf{I}^N = [-1/2, 1/2]^N$ of R^N, $\mathbf{y}_i$ coinciding with a vertex of the hypercube). Let $\mathbf{T}$ be a transformation consisting of a collection of $P \geq m+1$ affine maps of the form

$$\mathbf{T} = (T_1, T_2, \ldots, T_P) \; : \; T_j\mathbf{x} \equiv rR_j.\mathbf{x} + \mathbf{t}_j \, , \; j = 1, 2, \ldots, P. \tag{2.1}$$

where each component map T_j, carries a point $\mathbf{x} \in \mathsf{I}^N$ into a new point $\mathbf{x}_j = T_j\mathbf{x} \in \mathsf{I}^N$. The scaling factor r obeys $|r| < 1$ (same factor for all P maps); R_j is a rotation matrix (involving $N(N-1)/2$ free angular parameters for each map) and a translation vector (N free displacement parameters for each map), with the constraints

$$T_j\mathbf{y}_j = \mathbf{y}_j \quad , \quad j = 1, 2, \ldots, m+1. \tag{2.1.a}$$

The transformation $\mathbf{T}$ thus involves $f = (N+1)NP/2 - N(m+1) + 1$ free parameters (r, θ (rotation angles), $\mathbf{t}$ (translation coefficients)). The action on B of an individual map T_j produces a set

$$T_jB = \{x_j \in R^N \,|\, \forall \mathbf{x} \in B \Rightarrow \mathbf{x}_j = T_j\mathbf{x}\}. \tag{2.2}$$

The effect of the transformation $\mathbf{T}$ on B is finally defined by

$$\mathbf{T}B \equiv T_1B \cup T_2B \cup \ldots \cup T_PB = B_1 \tag{2.3}$$

Thus $\mathbf{T}$ generates a new set $\mathsf{B}_1^m \subset \mathsf{I}^N$ of same topological dimension m as the original set B^m. (3) Consider finally the iterative scheme

$$B_k = \mathbf{T}B_{k-1} \quad , \quad k = 1, 2, \ldots \quad \text{with} \quad B_o \equiv B. \tag{2.4}$$

In Mandelbrot's terminology the sets B_k are *prefractals*. Since each component map T_j is a contraction, the iterations (2.4) converge. The limit

$$\lim_{k \to \infty} B_k = H, \text{ with } H = \mathbf{T}H, \tag{2.5}$$

(invariant set of the transformation $\mathbf{T}$), is a *deterministic hierarchical set* specified by $(\mathbf{T}(P; r, \theta, \mathbf{t}), \mathsf{B})$. Geometrically the definition of H as an invariant of the iterative transformation $\mathbf{T}$ implies that this structure has the property that a linear magnification by a factor $\lambda = 1/r$ of one among the P 'pieces' $T_j\mathsf{H}$ of $\mathbf{T}\mathsf{H}$ is congruent to $\mathbf{T}\mathsf{H}$ (=H). Therefore, if $\mathcal{M}(\mathsf{H})$ is the $F_{\mathcal{M}}$-measure of the set H, then

[7] Recipes for generating fractal sets are found in Voss (1985, 1988) and Saupe (1988) (based on fractional Brownian motions and related techniques), and in Barnsley (1986, 1988a, 1988b) (IFS).

$$\mathcal{M}(H) = \mathcal{M}(\mathbf{T}H) = P\,\mathcal{M}(T_j H) = P\,r^{F_M}\,\mathcal{M}(H), \tag{2.6}$$

where the last equality follows from $T_j\mathsf{H}$ being a rescaled version of H (cf. 1.19). The second equality holds provided that the P pieces do not mutually intersect; this requires that for any pair $T_i \neq T_j$

$$T_i B_k \cap T_j B_k = \emptyset, \text{ at any iterative step } \quad k = 1, 2, \ldots \tag{2.6.a}$$

Hence the hierarchical set $(\mathbf{T}(P; r, \theta, \mathbf{t}), \mathsf{B})$ has the mass (similarity) dimension

$$F_M = \ln P / \ln \lambda = F_s, \tag{2.7}$$

(number of pieces P = mass ratio μ; Eq. 1.20). Note that in the presence of self-intersections of the resulting set H, Eq. (2.6) overestimates the 'mass' of the set so that the general expression for the similarity dimension becomes

$$F_s \leq \ln P / \ln \lambda \leq N. \tag{2.7.a}$$

Among the f free parameters of the transformation $\mathbf{T}$ only the linear scaling parameter, r, and the number of affine maps, P, enter the dimension. The general visual aspect of a limit set depends on the precise transformation $\mathbf{T}$. Accordingly, the dimension F_s is an incomplete indicator of the geometrical aspect of the set.

Instead of a single transformation $\mathbf{T}$ take a family $\mathbf{T}(s)$ depending on discrete or continuous parameters s, with associated probability density $p(s)$, the scaling parameter r and the number of individual maps P being kept constant. Then we define a ***statistical hierarchical set*** as the limit set of iteration (2.4), where at each step a transformation $\mathbf{T}(s)$ is selected according to the probability distribution $p(s)$. The mass dimension of the set is again given by Eq. (2.7).

2.2. Collection of Fractal Dimensions and Topological Indices

We first observe that while a priori, the basic sets B^m can be the origin O ($m = 0$), the unit line segment ($m = 1$), the triangle ($m = 2$), etc, the limit set (2.5) is independent of the dimension m of the basic set chosen. For if T_j^c denotes the map T_j with the translation terms discarded, then it follows from $r < 1$ that for any B 'centred' at the origin

$$\lim_{k\to\infty} (T_j^c)^k B = O \quad , \quad j = 1, 2, \ldots, P. \tag{2.8}$$

Geometrically this means that the limit set H is made up of an infinite number of individual points which are translations of the origin (2.8). The asymptotic set H is the counterpart of an *attractor* for the transformation $\mathbf{T}$. We carry out the following discussion for an embedding space R^2 only. It is then convenient to choose as the basic set the unit line segment B^1 on the x axis, centred at the origin. The hierarchical set $(\mathbf{T}(P; r, \theta, \mathbf{t}), \mathsf{B})$ is geometrically fully specified by its first iterate B^1. The latter defines a graph of P edges and v vertices, subject to 2 metric conditions : (a) one pair of vertices must be at a distance 1 from each other; (b) for any vertex V there is another vertex V' at a distance r from V. After K

iterations (Eq. 2.4,5), the prefractal B_K is given by a collection of P^K segments of length r^K each.

(A) Global dimensions < 1: select a value P and choose r such that $Pr < 1$. Then the first iterate B_1 can have no line segment of length ≥ 1, and cannot be connected. The resulting fractal is a *fractal dust*, characterised by a single dimension, namely the global mass dimension, in the range 0 to 1.

(B) Global similarity dimensions ≥ 1: (1) We can always select the parameters $(r, \theta, \mathbf{t})$ such that the first iterate is connected and traces out a *single* non-self-intersecting broken line (path), L made of P segments (topology of the interval I^1) joining the two endpoints of the initial segment:

$$B_1 = L(P). \tag{2.9}$$

The length of B_1 ($= P/\lambda$) then exceeds the length of B (=1). Assuming that condition (2.6.a) holds, any iterate B_K is again a path $\mathsf{L}(P^K)$; the resulting limit set H (Eq. 2.5) is a *filament*, characterised by a *single* fractal dimension, namely the global similarity dimension (2.7) (*filament dimension*).

(1.1) Substitute to the segment $\mathsf{L}(P)$ a closed non-self-intersecting polygon of P segments (*P-cycle*), $\mathsf{C}(P)$, so that the first iterate B_1 becomes a topological circle (containing $\mathbf{y}_1$ and $\mathbf{y}_2$). This topology requires $Pr > 2$. Under condition (2.6.a) the prefractal B^1_K then defines a connected loop of beads. The corresponding limit set (2.5) is a *hierarchical loop* characterised by a similarity dimension (Eq. 2.7) (*loop dimension*). The original loop $\mathsf{C}(P)$ is the union of two filaments, $\mathsf{C}(P) = \mathsf{L}(P_1) \cup \mathsf{L}(P_2)$, $P_1 + P_2 = P$. Therefore with the hierarchical loop structure we attach two independent filament dimensions, $\ln(P_i)/\ln\lambda$, $i = 1, 2$.

(1.2) Substitute for the path $\mathsf{L}(P)$ of construction (1) a *topological tree*, T, made up of a non-self-intersecting path of P' segments, $\mathsf{L}(P')$, (joining $\mathbf{y}_1, \mathbf{y}_2$) and b branches, $\mathsf{b}(P_i)$, of $P_i, i = 1, 2, \ldots, b$ segments respectively, attached to $\mathsf{L}(P')$, $P = P' + P_1 + P_2 + \ldots + P_b$:

$$B_1 = L(P') \cup b(P_1) \cup \ldots \cup b(P_b) \equiv T(P', P_1, \ldots P_b). \tag{2.10}$$

Under condition (2.6.a) the prefractal B^1_K then defines a connected arborescent structure. The limit set is a *hierarchical tree*. As independent dimensions we choose the global similarity dimension of this hierarchy (Eq. 2.7)(*tree dimension*); a second dimension is attached with the completely pruned tree, obtained by cutting off all branches; the pruned set, a filament generated from $\mathsf{L}(P')$, has a filament dimension

$$\ln P'/\ln\lambda = F'_s. \tag{2.11}$$

By partial prunings of all branches but $\mathsf{b}(P_i)$ we obtain arborescent subsets of subtree dimensions

$$\ln(P' + P_i)/\ln Q = F_s^{(i)}, i = 1, 2, \ldots, b. \tag{2.12}$$

Observe that if

$$P = P' + P_1 + P_2 + \ldots + P_c, \tag{2.13}$$

then we can introduce the following dimensions

$$F = \ln P/\ln\lambda\ ;\ F' = \ln P'/\ln\lambda\ ;\ F_i = \ln(P'+P_i)/\ln\lambda\ ,\ i = 1,\ldots,c, \quad (2.13.a)$$

(F, *global* dimension; F', a *special* dimension; and F_i, c *partial* dimensions). Then one dimension can be expressed in the (c+1) remaining dimensions. Thus

$$F = \ln[(\lambda^{F_1} - \lambda^{F'}) + (\lambda^{F_2} - \lambda^{F'}) + \ldots + (\lambda^{F_c} - \lambda^{F'}) + \lambda^{F'}]/\ln\lambda, \quad (2.14)$$

$$F' = \ln\{[(\lambda^{F_1} + \lambda^{F_2} + \ldots + \lambda^{F_c}) - \lambda^{F}]/(c-1)\}/\ln\lambda. \quad (2.14.a)$$

In a similar fashion we can generate *loops with attached branches*, or *trees with attached loops*, with corresponding loop, subtree and filament dimensions.

(2) Consider next the alternatives of *disconnected* iterates B_1. Choose the parameters (r,θ,t) such that the first iterate is made up of $1+f$ disconnected paths, a first non-self-intersecting path being formed by P' segments, $L(P')$, joining y_1, y_2, and f additional non-self-intersecting broken lines , $l(P_i)$, $i = 1,2,\ldots,f$, formed by P_i segments respectively, with $P = P' + P_1 + \ldots + P_f$

$$B_1 = L(P') \cup l(P_1) \cup \ldots \cup l(P_f). \quad (2.15)$$

Under condition (2.6.a) the prefractal B_K is then the union of a path $L(P'^K)$ and a hierarchy of f disconnected paths surrounded by disconnected paths, etc. The limit set (2.5) is the union of a main filament, and an infinite hierarchy of filaments surrounded by filaments (*chain of filaments*). We have again $(1+p)$ obvious independent fractal dimensions (Eqs. 2.13.a): the global similarity dimension of the hierarchy (*filament chain dimension*); the similarity dimension of the main filament $\ln P'/\ln\lambda$ (*filament dimension*); and the filament subchain dimensions corresponding to the subchains of filaments, of the chain obtained by eliminating all segments but $l(P_i)$, $i = 1,2,\ldots,f$.

(2.1) Construction (2) can be complicated by attaching b branches to the main path $L(P')$, to form a tree $T(P', P_1', \ldots, P_b')$, and to the auxiliary segments $l(P_i)$ each time b_i branches to form again trees $t(P_i, P_{i,1}, P_{i,2}, \ldots, P_{i,b_i})$, to obtain a *chain of trees*. We have

$$P = P' + (P_1' + \ldots + P_b') + (P_1 + P_{1,1} + \ldots + P_{1,b_1}) + \ldots + (P_f + P_{f,1} + \ldots + P_{f,b_f}), \quad (2.16)$$

defining the global dimension (2.13.a) of the overall tree; as the special dimension we can take the filament dimension describing the pruned main tree; etc.

(2.2) A further way of complicating construction (2) amounts to transform the auxiliary filaments of the first iterate into c (non-self-intersecting) cycles, $c(P_i)$ formed by P_i segments, $i = 1,2,\ldots,c$. Then $P = P' + P_1 + P_2 + \ldots + P_c$

$$B_1 = L(P') \cup c(P_1) \cup \ldots \cup c(P_c). \quad (2.17)$$

This hierarchical set is a *chain of islands*, union of a filament (coastline), and of a hierarchy of islands. The global similarity dimension of the hierarchy describes the border of the islands plus coastline (*island chain dimension*); the special dimension is again a filament dimension, (similarity dimension of the filament, and of the boundary of any individual island of the chain). The partial dimensions are island subchain dimensions which correspond to the subchains obtained by suppressing all polygons but one, $\mathsf{c}(P_i)$ in the first iterate.

(2.3) Modify construction (2.2) by the substitution $\mathsf{L}(P') \to \mathsf{C}(P')$ in the first iterate, $\mathsf{C}(P')$ being a cycle formed by $P' = P'_1 + P'_2$ segments (containing $\mathbf{y}_1, \mathbf{y}_2$), with the c cycles, $\mathsf{c}(P_i)$ formed by P_i segments, $i = 1, 2, \ldots, c$ being preserved. Then $P = P'_1 + P'_2 + P_1 + P_2 + \ldots + P_c$

$$B_1 = C(P') \cup c(P_1) \cup \ldots \cup c(P_c). \tag{2.18}$$

The hierarchical set is reminiscent of a two dimensional chain of soap bubbles, (*chain of loops*), or hierarchy of hierarchical loops (cf. 1.2).The global dimension describes the border of the chain of loops (*loop chain dimension*); the special dimensions $\ln P'_j / \ln \lambda$, $j = 1, 2$ are filament dimensions; $\ln P' / \ln \lambda$ is a loop dimension. The partial dimensions $\ln(P'_j + P_i) / \ln \lambda$ are island subchain dimensions, while the partial dimensions $\ln P' / \ln \lambda$ are loop subchain dimensions.

2.3. A Symbolic Notation for Hierarchical Sets in the Plane

With an alphabet of ***signs*** $i, j, \ldots$, elementary ***symbols*** of pairs of signs, (i, j), and a ***product*** operation between symbols we construct a collection of symbols Σ by the following rules.

(a) (i, j), for any pair of signs i, j, is a symbol Σ of length 1;

(b) $(i, j) = (j, i)$, for any pair of signs;

(c) if Σ' and Σ'' are symbols of length P' and P'' respectively, then the product $\Sigma' \Sigma''$ is a symbol Σ of length $P = (P' + P'')$; Σ' and Σ'' are ***divisors*** or subsymbols of Σ;

(d) $\Sigma' \Sigma'' = \Sigma'' \Sigma' = \Sigma$, for any pair of symbols.

Any symbol of length P can therefore be expanded as a product of P symbols of length 1. We use the following terminology (Temperley 1981).

(a') A sign i is an *endpoint* of a symbol Σ if it enters in one and only one of its divisors of length 1;

(b') the symbols $\Sigma', \Sigma'', \Sigma''', \ldots$ are *disconnected* if they have no sign i in common (belonging to all of the symbols); the number of *connected pieces*, n_c , of a symbol Σ is the number of its disconnected divisors;

(c') a symbol is a ***self-avoiding n-path***, π, if it can be rearranged by properties (a)-(d) in the form

$$(a, i_1)(i_1, i_2) \ldots (i_{n-1}, b) = \pi, \tag{2.19}$$

i.e. any sign i_k, $k = 1, 2, \ldots, n-1$ occurs in exactly 2 symbols of length 1; the unpaired signs a , b are the endpoints of the path;

(d') a symbol is an ***n-cycle***, γ , if it can be rearranged in the form of an n-path (2.19) whose endpoints are identical $a \equiv b = i_o$;

(e') a symbol is an ***r-star*** (r-rayed star, $r > 2$), σ, if it has length r and if it can be rearranged as

$$(a, i_1)(a, i_2) \ldots (a, i_r) = \sigma, \tag{2.20}$$

i.e. any pair of length-1 divisors of (2.20) has exactly one sign a in common; sign a is the *root*, and r is the *valency* of a.

The number of independent cycles of a symbol, χ, is given by

$$\chi = P - v + n_c, \tag{2.21}$$

(v, number of distinct signs; Temperley 1981).

With a geometric point in the plane associate a sign i; to a line segment of unit length joining two points symbolised by i and j there corresponds the pair symbol (i, j). The union of two geometric sets of associated symbols Σ' and Σ'' corresponds to the symbolic product. With these conventions, the geometrical structure of any first iterate B_1 is encoded in a symbol Σ of the above form. We shall use the signs 0 and 1 for the invariant points of the two affine maps T_1 and T_2 (Eq. 2.1.b). Conversely, any symbol generated by the rules (a-d) can be interpreted as a first iterate B_1. Since the product of any symbol by any other symbol produces again a symbol (rule c) the notation implies that we can generate *arbitrarily complex first iterates* (in the sense of Kolmogorov's algorithmic complexity, 1965). The hierarchical sets discussed in the literature so far (cf. Mandelbrot 1982) are all based on lowest complexity first iterates.

The algebraic notation extends to hierarchical sets in N-dimensional Euclidean embedding spaces, $(i_1, i_2, \ldots, i_N)$ (symbolising $(N-1)$-dimensional simplices) (a); (b) is to be replaced by the condition that any permutation of the labels i_k represents the same symbol; (c) and (d) are unchanged. For $N \geq 3$ a classification of the symbols is wanting (all 1-manifolds are topological circles, while there is a countable infinity of oriented 2-manifolds, namely the spheres of genus g S^2_g; $g = 0$, standard sphere; $g = 1$, torus; etc).

The symbol Σ ($\equiv \mathsf{B}_1$), of length P, has $(2^P - 1)$ different divisors (subsets of B_1, barring the empty set), of lengths $1, 2, \ldots,$ and P; there are $C^L_P = P!/[L!(P-L)!]$ divisors of length L, with fractal dimensions $\ln L/\ln \lambda$, regardless of the topology of B_1. This makes it clear that the dimensions of the totality of subsets of B_1 provide no information at all on the distinguishing features of hierarchical sets of same dimensions. Visual differences are related to topological characteristics of the first iterate, (occurrence of paths, cycles and stars), as well as to dimensions of particular subsets. In order to isolate the latter symbolically, a systematic factorisation procedure of the symbol of the first iterate is adopted.

Factor first Σ into *disconnected divisors*

$$\Sigma = \Sigma' \Sigma'' \ldots \Sigma^{(n_c)}, \tag{2.22}$$

with Σ' the main symbol containing both signs 0 and 1 (which may be empty); the number n_c is the topological connectivity index; associated with Σ' we have the special dimension, $\ln P'/\ln \lambda$ (Eq. 2.13.a). This factorisation provides the main topological characterisation of the hierarchical set.

If the main symbol is empty (no filament joining 0 and 1), and if no divisor $\Sigma^{(i)}$ has a length $P^{(i)}$ such that $P^{(i)} r \geq 1$, then the hierarchical set is a *fractal dust*; graphical representations indicate that often the global dimension together with the partial dimensions $\ln P^{(i)}/\ln\lambda$ are accurate visual identifiers, regardless of the topology of the subsets $\Sigma^{(i)}$. If one divisor $\Sigma^{(j)}$ has a length such that $P^{(j)}r \geq 1$, then a scar of the topology of the latter remains noticeable in the hierarchical set, and a specification of the topology of $\Sigma^{(j)}$ is required (cycle, star, etc); dimensions of the subsets of $\Sigma^{(j)}$ should be indicated as well.

If Σ' is not empty, then factor the latter sign into main paths $\pi'_1,\ldots,\pi'_p$ of shortest lengths joining 0 and 1; and into secondary paths $\pi''_1,\ldots,\pi''_s$ joining signs of the main paths, or of the secondary paths; and remaining symbols $\Sigma'_1\Sigma'_2\ldots\Sigma''_r$, rooted at a sign of a main or a secondary path (arborescences)

$$\Sigma' = \pi'_1\pi'_2\ldots\pi'_p\pi''_1\pi''_2\ldots\pi''_s\Sigma'_1\Sigma'_2\ldots\Sigma'_r, \tag{2.23}$$

The p main paths determine p independent filament dimensions (≥ 1). If one filament dimension is $\equiv 1$, then the construction of the hierarchical set implies that $\mathsf{B}_1 \subset \mathsf{H}$. If $p = 1$, $s = 0$, $r = 0$ and if moreover $\Sigma \equiv \Sigma'$ (Eq. 2.22), then the corresponding hierarchies are Von Koch curves (cf. Mandelbrot 1982 for illustrations) whose visual structure is fully characterised by the single global dimension $\ln P/\ln\lambda$ (a filament dimension).

If $p = 1$, $s = 0$ but $r \neq 0$ (presence of stars), we obtain, for $\Sigma \equiv \Sigma'$, Mandelbrot-Given curves (1984) as the simplest representatives. In addition to the global dimension, the dimension of the main path π' determines the visual effect of the hierarchy.

If $p > 1$, the number of independent main cycles of Σ' ($\equiv$ number of independent loop dimensions) is $c = p - 1$ (Eq. 2.21). That the visual effect then depends on the loop dimensions, as well as on the filament dimensions, is made clear by a comparison of the following examples (take $r = 1/3$): (a) $\Sigma =$ (0,a) (a,b) (b,c) (c,1) (1,d) (d,e) (e,f) (f,0): main cycle of length 8 decomposed into two main paths $\pi'_1 =$ (0,a) (a,b) (b,c) (c,1) and $\pi'_2 =$ (1,d) (d,e) (e,f) (f,0); global dimension = loop dimension = ln 8/ln 3 $\approx$ 1.89; two equal filament dimensions = ln 4/ln 3 $\approx$ 1.26; (b) $\Sigma =$ (0,i) (i,j) (j,k) (k,l) (l,1) (1,m) (m,n) (n,0): main cycle decomposed into two main paths (0,i) (i,j) (j,k) (k,l) (l,1) and (1,m) (m,n) (n,0); global dimension = loop dimension = ln 8/ln 3; two filament dimensions = ln 5/ln 3 $\approx$ 1.46 and 1.

Finally, factor each secondary symbol $\Sigma^{(i)}$ ($\Sigma'',\ldots,\Sigma^{(n_\circ)}$) into one longest path, $\pi^{(i)}$, or longest cycle, $\gamma^{(i)}$, respectively, and secondary paths and arborescences (cf. Eq.2.22); the former subsets produce visually noticeable dimensions.

Given a picture of a set, standard procedures of image analysis (Serra 1988) allow one to prune the image, to isolate connected parts, etc, i.e. to generate subsets which – on condition that the image is a hierarchical set – arise from noticeable subsets of the first iterate. Therefore we can directly measure a *collection of dimensions*, the global dimension, F, and dimensions $F' > F'' > \ldots > F^{(k)}$ of the noticeable subsets of the observed set. A hierarchical set specified by these dimensions exists only if the system of $(k+1)$ equations

$$\ln P = F\ln\lambda \quad ; \quad \ln P' = F'\ln\lambda \quad ;\ldots; \quad \ln P^{(k)} = F^{(k)}\ln\lambda \tag{2.24}$$

in the $(k+1)$ unknown integers $P, P'\ldots P^{(k)}$ and in the unknown $\ln\lambda$ admits of a solution. Since the dimensions are measured with a finite precision, there is a collection of smallest integers $P_o, P'_o, \ldots$, (and with λ an integer), satisfying these equations. Hence any hierarchical set defined by a first iterate of global length P_o is always a *formal* solution to this problem. In order to be geometrically acceptable, the first iterate must be constructed to have noticeable subsets of length P_o, P'_o etc. Typically, in actual image analysis only 2 or 3 dimensions are directly measured; the geometric problem then has a (non-unique) solution. For instance, if the global dimension, F, and the dimension of the pruned set, F', are measured, and if the set is seen to be connected, then the first iterate is given by the symbol $\Sigma' = \pi\Sigma'_1\Sigma'_2 \ldots \Sigma'_r$, (main path with branches), of total length P_o, with the main path having length P'_o; the r branches have a total length $P_o - P'_o$ ($\geq r$), so that neither the total number of the latter, nor their topology is specified. To simulate the observed set, one can construct a statistical hierarchical set in which different realisations of the first iterate compatible with the algebraic constraints (2.24) define different representatives of the statistical set of transformations $\mathbf{T}(s)$.

Power law distributions, such as the number of linear oscillation frequencies *less than* a given frequency, can be interpreted in terms of pseudo-dimensions. We mention here that a second class of power law distributions, such as the number of stars *greater than* a given mass, the number of impact craters of diameter *greater than* a given diameter, can be related to the geometry of hierarchical sets. This is made clear by the elementary examples discussed in Perdang (1990b) (first iterate of hierarchy given by the symbol $\Sigma = \Sigma'\Sigma''$, Σ' a single path, Σ'' a single cycle; the diameters D of the cycle then obey a law $N(D) = KD^{-a}$, $a \equiv F_s$ the global similarity dimension of the hierarchical set).

3. Dynamical Generation of Fractals

While the algebraic construction (2.4) of hierarchical sets does not mention time explicitly, we can interpret step k in the iteration as a discrete *timestep* t. The algebraic model then transforms into an evolutionary iteration

$$B_{t+1} = \mathbf{T}B_t, \quad \text{with} \quad B_o = B \tag{3.1}$$

(B, initial geometrical set). If we regard the sets as *points* of an abstract space, then the hierarchical set is a *fixed point* of the transformation $\mathbf{T}$ in this space. By successive reinterpretations of this space, Eq. (3.1) can be viewed as representing a *cellular automaton*, a *standard iterative map*, or a *differential system*.

3.1. Deterministic Cellular Automata

An n-dimensional cellular automaton (nD CA) is an n-dimensional lattice of cells identified by integer Cartesian coordinates $i = (i_1, i_2, \ldots, i_n)$, with a discrete field variable $Y(i,t)$ attached with each cell, or site, i, at each discrete time $t = \ldots, -1, 0, +1, \ldots$,

$$Y(i,t) \in Z_k = 0, 1, 2, \ldots, k-1 \text{ (set of integers modulo } k). \tag{3.2}$$

The lattice space is either infinite ($i_j = \ldots, -1, 0, 1, \ldots$), or toroidal ($i_j \in \mathsf{Z}_s$, s possibly different for different coordinates i_j). The *state* of the automaton at time t is the collection of values of the field variable Y over all sites

$$\mathbf{Y}(t) = \{\ldots, Y(i,t), Y(i',t), Y(i'',t), \ldots\}. \tag{3.3}$$

Take for definiteness $k = 2$ (Eq. 3.2). The array $\mathbf{Y}$ then identifies a geometric set Y of cells in state 1 (a subset of the lattice space). The evolution in time of the CA is specified by a 'neighbourhood' $\mathsf{n}(i)$ selected for each cell i , and by a 'local evolution rule' R which fixes the field at site i and time $t+1$ in terms of the field at the sites $i' \in \mathsf{n}\ (i)$ at time t

$$Y(i,t+1) = R(Y(i,t), Y(i',t), Y(i'',t), \ldots) \quad , \quad i, i', i'', \ldots \in n(i). \tag{3.4}$$

(*deterministic* CA). A second alternative consists in randomly choosing at each time, or each time and site, a local rule R among a collection $R_1, R_2, \ldots, R_S$ with associated probabilities $p_1, p_2, \ldots, p_S$ (*stochastic* CA). The evolution can be summarised by an equation of type (3.1)

$$Y_{t+1} = \mathbf{R} Y_t, \tag{3.5}$$

(Y_t, geometric set defining the state of the automaton at time t; $\mathbf{R}$, geometric operator; $\mathsf{Y}_o = \mathsf{Y}$, initial geometric set of the lattice). The formal analogy between iterations (3.5) and (3.1) is only superficial. The operator $\mathbf{T}$ transforms the unit cube into itself, while the cellular automaton operator $\mathbf{R}$ contains in general translation terms which produce a *propagation* in different directions, of maximum velocity V depending on the neighbourhood. However, if we rescale Y_t, setting $\mathsf{V}_t = \mathsf{Y}_t/(t+1)$, $t = 0, 1, \ldots$, then the geometric configuration V_t remains confined in the spatial region Y. Moreover, at each step V_t typically acquires an exponentially higher degree of structure, so that V_t converges to a hierarchical geometry H (in an infinite lattice). Hence, if we can simulate the dynamics of a phenomenon by a CA model, then the model produces generically fractal structures in space. This conclusion is supported by systematic numerical studies of 1D and 2D CA's defined over an infinite lattice space. The behaviour of 1D CA's with field $Y \in \mathsf{Z}_2 = \{0,1\}$ and neighbourhoods of site i consisting of i itself, its r 'left' ($i-1, \ldots, i-r$) and r 'right' neighbours ($i+1, \ldots, i+r$) have been investigated in space-time for all deterministic rules [8] (Wolfram 1983, 1984, 1985, Martin et al. 1984; cf. also Manneville et al. 1989). For $r = 1$, and initial condition such that a single cell is in state 1, all other cells being in a quiescent state 0, the collection of excited states 1 shown over a discrete space-time lattice exhibits 3 classes of patterns. Class 1: equilibrium state; Class 2: periodic pattern; Class 3: self-similar pattern in the 2D lattice space-time, in the following sense: to the actual set $(\mathsf{Y}_o, \mathsf{Y}_1, \ldots, \mathsf{Y}_t) \equiv \mathsf{U}_t$, ($\mathsf{Y}_o$, Y_1, ... constant time slices) substitute $\mathsf{W}_t = \mathsf{U}_t/(1+t)$; for t tending to infinity this set remains confined to the initial set, and

[8] For a finite neighbourhood, and a finite number of states the number of deterministic rules is finite and can be explored by numerical experiments.

it has structure at any scale. For $r > 1$ a fourth type of behaviour is observed (Class 4): complex localised arborescent structures. For random initial conditions the same 4 classes of behaviour are found. Wolfram's statistical experiments show that with increasing r the frequency of Class 3 behaviour tends to dominate. [9] The same observation holds if the number of states k is increased.

The prototype of a hierarchy in a 1D CA is produced by the 'modulo 2 rule'

$$Y(i,t+1) = Y(i-1,t) + Y(i+1,t) \mod 2 . \tag{3.6}$$

For a single initially excited cell the rule is equivalent to the relation between the binomial coefficients $C_{i+1}^{k+1} = C_i^k + C_i^{k+1} \mod 2$ (Pascal's triangle modulo 2); the space-time pattern is therefore a self-similar structure whose similarity dimension is read off from the first stages of the iteration $F_s = \ln 3/\ln 2 \approx 1.58$ (Eq. 1.20). A random line in space-time intersects this hierarchical set to produce again a fractal set of mass dimension $F_s - 1$ (Eq. 1.16); if the space or time sections of these fractals were generic, we would see again typically fractal sets of dimension $F_S \equiv F_T = F - 1 = 0.58$. The space and time axes have privileged directions here, so that this estimate is not applicable. For a similar investigation of 2D CA's see Packard and Wolfram (1985). For 2-state nD CA's with Von Neumann neighbourhood (nearest neighbour cells) self-similar space-time patterns of similarity dimensions $\ln(2n+1)/\ln 2$ and $\{\ln(n) + ln[1 + (1 + 4n - 1)^{1/2}]\}/\ln 2$ are obtained for the modulo 2 evolution rule with cell i included or excluded respectively.

As a trivial qualitative model for the evolution of luminous matter in the universe take a 2-state deterministic 3D CA (empty space or dark matter 0; and luminous matter 1) with rule $0 \rightarrow 1$ iff the Von Neumann neighbourhood contains an odd number of cells of luminous matter (standard modulo 2 rule). Hence according to the previous results the luminous matter distributes over a space-time fractal of dimension 2.81; for constant time slices we have the estimate $F = 1.8$. The dynamics of this model provides a qualitative interpretation for the observed fractal-like distribution of the galaxies or clusters of galaxies (exponent of the two-point correlation function of galaxies $\gamma = 1.8$; Peebles 1980, Davis and Peebles 1983; Bahcall and Soneira 1983; hence the observed $F = n - \gamma = 1.2$ for $n = 3$, which differs quantitatively from the trivial model). A slightly more detailed CA has been studied by Bak and Chen (1989). A 3-species universe is modelled by a 3-state deterministic 2D or 3D CA (empty space 0, inert matter 1, active matter 2) with Von Neumann neighbourhood, with rules (a) $2 \rightarrow 1$; (b) $1 \rightarrow 0$ iff single active nearest neighbour; (c) $0 \rightarrow 2$ iff one active and one passive nearest neighbour on opposite sides of the cell. A stationary state is found in which the number of active sites in a 2- or 3-ball centred at an active site scales with the radius R like R^F, with $F = 1.6 \pm 0.2$ and 1.7 ± 0.2 in the 2D and 3D case respectively.

[9] For $r = 3, k = 2$ the relative frequency of class 3 behaviour is 0.73.

3.2. Stochastic Cellular Automata

The fractal structures connected with *site percolation* (formation of a fractal percolation cluster; cluster scaling laws; Essam 1980, Stauffer 1985, 1986; Rapaport 1985, Herrmann 1986) and *diffusion* (fractal diffusion hull; relation of the scaling exponents of percolation with the dimension of the hull and the cluster; Sapoval et al. 1985, 1989) as illustrations of a stochastic CA have been discussed in Perdang (1990b), where the relevance of these mechanisms for the interpretation of the shape of molecular clouds has been emphasised.

Two dynamical models of the galaxy distribution have been presented in the form of stochastic CA's: Schulman and Seiden (1986) adopt a 2-state 3D CA (empty cell 0; cell occupied by a galaxy 1) obeying the rules : $0 \rightarrow 1$, with a probability p_1 per timestep (spontaneous galaxy formation); and $0 \rightarrow 1$, (probability p_2) if a cell of the neighbourhood is occupied (Ostriker-Cowie 1981 enhanced galaxy formation). The formation processes are stopped when the occupation probability p (number of activated cells/total number of cells) comes close to the *percolation threshold* p_c (the critical value at which the percolation cluster develops). The pair-correlation exponent then obeys $\gamma \approx 1$, i.e. $F \approx 2$. Note that the quantitative agreement with observation is worse than for the trivial deterministic models.

Vicsek and Szalay (1987) take a 2-state 2D CA with Von Neumann neighbourhood, in which a continuous field variable y defines the standard 2-state CA variable Y (empty cell 0; occupied cell 1). The evolution rule is expressed in the continuous field and in a free parameter as follows: $y(t+1)$ in cell i is equal to the average of the $y(t)$'s over the neighbourhood, plus a random correction r ($= -1$ or $+1$, with equal probability); then the state of cell i is updated to 0 or 1 depending on whether $y \leq A$ or $> A$, with A a free threshold parameter (density induced galaxy formation). The model generates a fractal galaxy distribution of a dimension strongly dependent on the threshold parameter; for $A \approx 4.75$ the observed value $\gamma \approx 1.8$ is reproduced.

Lejeune and Perdang (1990a, 1991) adopt a 5-state 2D stochastic CA model for the evolution of the amount of gas, protostars, main sequence stars, and red giants in a galactic environment. The model extends the forest-fire model of star formation (cf. the CA of cosmic evolution) of Gerola and Seiden (1978), Seiden and Gerola (1982) and Schulman and Seiden (1982, 1983). The evolution rules are of the form: state $k \rightarrow$ state k', with probability $p_{kk'}$ per timestep (given by standard stellar evolution theory); state $k \rightarrow$ state k'', if favourable conditions in the cell neighbourhood are satisfied, with probability $p_{kk''}$ (enhanced evolution); in addition mechanical rules simulating galactic rotation and mixing are implemented. It is found that for a broad range of the free parameters the system evolves towards a statistically stationary state with a spatial distribution of the red giants of mass dimension ≈ 1.7 and correlation dimension ≈ 1.6 (cf. 1.29.b). The theoretical possibility of spatially fractal distributions of stars has been envisaged by Ikeuchi and Yoshioka (1989) (fractal distribution of supernova remnants). The numerical experiments also indicate that ranges of the probabilities exist over which the different global populations undergo approximately periodic variations in time, with a period of 4 timesteps and a spatial coherence extending over ≈ 50 cells.

3.3. Iterative Maps

We generate an extension of a CA by regarding the field variable as *continuous*; simultaneously the homogeneity of the CA lattice space, expressed by evolution rules which are the same for all cells, may be relaxed. The CA then transforms into an *Iterative Map* (IM) whose state vector $\mathbf{Y}(t)$ (Eq. 3.3), of c components (= number of cells of the lattice, finite or infinite), regarded as defined over a c-dimensional manifold D^c (*state space* or *phase space*), obeys the evolution rule

$$\mathbf{Y}(t+1) = \mathbf{R}(\mathbf{e};\mathbf{Y}(t)) \tag{3.7}$$

($\mathbf{R}$, a collection of c single-valued functions R_i depending continuously on a set of parameters $\mathbf{e} = (e_1, e_2, \dots, e_S)$, $\mathbf{e} \in \mathsf{S}^S$). For fixed parameters the IM is *deterministic*; for parameters obeying a probability distribution such that at each timestep a rule $\mathbf{R}(\mathbf{e};\dots)$ is selected with probability $p(\mathbf{e})d\mathbf{e}$, the IM is *stochastic*. From an extrapolation of Wolfram's experiments of generic occurrence of hierarchical sets in the discrete CA space-time we infer that fractals should also arise generically, and in the *majority* of IM's. A difference between CA and IM models we notice in practice is that typically the lattice space of the standard CA remains the physical configuration space (cf. the models of the previous section). IM models often come from abstract formulations, so that the lattice space has a less direct interpretation (space of expansion coefficients, etc); accordingly, we can have physically meaningful IM's with a very small number of 'sites' ($c = 1$ or 2). We concentrate here on the latter IM's, for which fractal structures are observed in the *state space*, in the form of *attractors*; fractals are also encountered in the space of initial conditions, and in the parameter space.

(a) The simplest nontrivial IM, obtained for $c = 1$, and evolution rule involving a lowest order nonlinearity (polynomial of degree 2) [10] is the *logistic map*, of manifold of definition (state space) $\mathsf{D}^1 = [0,1]$ and of evolution rule

$$Y(t+1) = R(e;Y(t)) \equiv e\,Y(t)\,(1 - Y(t)), \tag{3.8}$$

(May 1976, 1983; cf. also Barnsley 1988, Section 4.3). Simple algebra shows that as the parameter e is increased from 0 to 4 the time behaviour of the stable solution exhibits the sequence

$$Z_1 \to Z_2 \to Z_4 \to Z_8 \to \dots \to Z_{2^\infty} \tag{3.9}$$

(Z_n,integers modulo n : n-cyclic, or period-n behaviour), bifurcations occurring at critical values e_n of e obeying the universal ratio (Feigenbaum 1979)

$$\lim_{n\to\infty} (e_{n+1} - e_n)/(e_{n+2} - e_{n+1}) = \delta = 4.6692\dots \tag{3.10}$$

[10]For a linear evolution rule equations (3.7) can be solved analytically for any t; nonlinear evolution rules typically generate a number of terms increasing exponentially with discrete time; this exponential multiplication of complexity is the counterpart of the increase of geometric complexity of the iterative scheme (2.4).

The Z_{2^∞} behaviour (*chaotic time-behaviour*) is found in the range $e_\infty = 3.57\ldots$ to 4, which also contains very small windows over which all other periods are observed. The geometric support of the chaotic solution (*attractor*) is an uncountable collection of points – a Cantor dust – in the state space, of fractal dimension ≈ 0.538 (cf. Rasband 1990). Dynamically the chaotic solution is characterised by sensitive dependence on the initial conditions (positive Lyapunov index, cf. Devaney 1986).

For IM's with higher dimensional state spaces ($c > 1$) the state vector $\mathbf{Y}(t)$ is supported by the following attractors: a *fixed point* (Z_o), a discrete n-point *limit cycle* (Z_n) (period-n oscillation), a discrete *torus* (direct product of Z_n's of different n) (multiply periodic solution), or alternatively a *fractal (strange) attractor* (chaotic oscillation) in the state space.

(b) In general the state space is partitioned into *basins of attraction* $D^c(T)$, $T = 1, 2, \ldots$, such that if $\mathbf{Y}(t) \in D^c(T)$ at some instant t, then $\mathbf{Y}(t') \in D^c(T)$ at any later time t'; attached with each basin we have a unique attractor $A(T)$ determining the type of eventual time behaviour T (equilibrium, period-n, chaotic oscillation). The boundaries of the basins of attraction are typically fractal sets (cf. the computergraphics in Peitgen and Richter 1986).

(c) For a fixed initial condition $\mathbf{Y}_o$ the parameter space S^S is partitioned into basins of different types T of behaviour, $S^S(T)$. Again, the boundaries of the latter are generally fractal sets.

As a first astrophysical illustration we mention an IM simulating the *radial oscillations* of a star. Under approximations discussed in detail elsewhere (Perdang 1989; see also Perdang 1990a for a more formal derivation), a $c = 2$ IM is generated from the stellar oscillation equations in differential form, of evolution rule

$$Y(t+1) = [(1+a)Y(t) + Y(t)^2 + bY(t)^{*2} + c|Y(t)|^2]/(1+\epsilon|Y|^4), \quad (3.11)$$

($Y(t) = Y_1(t) + iY_2(t)$, $Y_1(t), Y_2(t)$, expansion coefficients in the eigenfunctions of the relevant modes). The real part a_r of the complex coefficient a is essentially the linear vibrational stability coefficient of the star. The remaining complex parameters b and c characterise the nonlinear coupling effects; ϵ is a small real coefficient. The numerical exploration of the geometry of the basins of different time-behaviour in the a, b, c parameter space has been made by plotting (a_r, b_r), and (a_r, c_r) sections of the basins (Perdang 1989, 1990a). These sections indicate that (1) basins of any period-n behaviour of nonzero area do exist; (2) under a regular increase of the instability parameter a_r the Feigenbaum scenario of period bifurcations (3.8) is encountered at constant nonlinearity parameters (provided that the latter lie in a given range); the elementary IM thus duplicates the qualitative results of the full hydrodynamic calculations (Kovacs and Buchler 1988) (a_r is related to the effective temperature); (3) basins of different time behaviour are separated by boundaries which appear as smooth at very low resolution; they seem to be entirely irregular at intermediate resolution; at very high resolution (step in $a, b, c \approx 10^{-6}$ or 10^{-7}) the boundary is found to be a direct product of a *Cantor set* and an interval; the fractal mass dimension F of the Cantor set is 0.68 to 0.77 (Perdang 1990a). This model is indicative that in the HR diagram the strips of different time behaviour are separated by fractal edges, rather than by locally straight lines.

More recently a *stochastic* version of this IM representative for stellar pulsations has been introduced to account for the interaction of the oscillations with convection (Perdang 1991). The effect of stochasticity is to transform the attractors of the ensemble-averaged deterministic IM into *fuzzy sets*. If two fuzzy attractors have a nonempty (fuzzy) intersection, then the oscillation exhibits a new kind of *intermittency* (cf. the illustrations in Perdang 1991).

As a second astronomical illustration of an IM, I mention a model proposed by Hénon (1988, 1989) to interpret the observed hierarchically structured organisation of Saturn's rings as revealed by the Pioneer and Voyager encounters (cf. Stone and Miner 1981; Goldreich and Tremaine 1982). The model is based on a simulation of a restricted 3 body problem (central planet and 2 test particles of small mass circling around the planet with initially slightly different radii, studied by Petit and Hénon 1986) by an inclined planar billard, with lower reflecting border made of two intersecting circles. A ball dropped from a fixed height and a horizontal distance a from the point of intersection of the circles either may (a) remain trapped between the two circles, bouncing back and forth indefinitely (formation of a ring); or (b) it may execute a finite number n of bounces and then get ejected (gap in the ring). The billard is equivalent to a 2D IM depending on the single parameter a; it is shown analytically that the set of a values leading to alternative (a) (trapped orbits) is indeed a Cantor set (Hénon 1988, 1989).

3.4. Fractals Connected with Differential Equations

The IM (3.7) has an associated ODE if the evolution rule is written

$$\mathbf{R}(\mathbf{e};\mathbf{Y}(t)) = \mathbf{Y}(t) + \tau\mathbf{F}(\mathbf{e};\mathbf{Y}(t)) \ : \ d/dt\,\mathbf{Y}(t) = \mathbf{F}(\mathbf{e};\mathbf{Y}(t)), \tag{3.12}$$

with the timestep $\tau \to 0$, and time being regarded as continuous. If in addition the lattice space becomes a standard Euclidean space (cellsize tending to zero), then the locality requirement of the evolution rules in the physical space introduces partial space derivatives; the IM then has an associated PDE,

$$\partial/\partial t\, Y(\mathbf{r},t) = \mathbf{H}(\mathbf{e}; Y(\mathbf{r},t), \partial/\partial\mathbf{r}\, Y(\mathbf{r},t), \ldots), \tag{3.13}$$

provided that the relevant limit processes exist. ODE's and PDE's thus belong to extended IM's involving *limit processes*, so that it is not guaranteed that a given property of a solution of an IM remains preserved in the solution of the associated ODE or PDE. As examples, the first order ODE associated with the logistic map (3.8) possesses only one type of attractor (fixed points); the second order ODE associated with the stellar pulsation IM (3.11) has 2 allowed types of attractors (fixed points, limit cycles); the period-n and the chaotic oscillations of the IM's are lost in the limit process. The continuous phase space of both an IM and its associated ODE is covered by a continuous network of field lines, of tangent $\mathbf{F}(\mathbf{e};\mathbf{Y})$ at point $\mathbf{Y}$; the continuity of the time variable in the ODE then requires the solutions themselves to be continuous functions of time, so that the trajectory $\mathbf{Y}(t)$ in the phase space continuously follows a field line. In contrast, in an IM the solution points $\mathbf{Y}(t_i)$, although constrained to lie on the field lines, are not subject to the continuity requirement; as a consequence the solution may jump irregularly from one point of the field line to another, thereby generating period-n or chaotic solutions. From the single-valuedness of the field lines it follows that the minimum dimension of the ODE phase space needed to produce chaotic behaviour is 3.

A different connection between ODE's and IM's is borne out by the ***Poincaré surface of section*** method (cf. for instance Mera 1987): a $(c-1)$-dimensional plane (surface of section) transverse to the family of trajectories in the c-dimensional phase space of an ODE determines a sequence of successive intersection points with each trajectory of the family; the temporal succession of points in the plane is then defined by a standard $(c-1)$-dimensional IM. Accordingly, c-dimensional ODE's $(c \geq 3)$ exhibit the same variety of attractors as $(c-1)$-dimensional IM's, namely fixed points, limit cycles, m-tori and strange attractors. In particular the existence of a strange attractor of fractal mass dimension F in the Poincaré section indicates a strange attractor of fractal dimension $F+1$ of the ODE. Conservative ODE's (*Hamiltonian systems*), in which $\partial/\partial\mathbf{Y} \bullet \mathbf{F}(\mathbf{e};\mathbf{Y}) \equiv 0$, have received particular attention (cf. Arnold 1976). Such systems have no attractors; the generic solutions are *multi-periodic* (carried by m-tori in the phase space, $c = 2m$) or *chaotic* (carried by 'exploded' tori which cover a finite 'volume' in the $(c-1)$-dimensional energy surface); the chaotic region in the phase space is interspersed with families of nested tori, surrounded by second generation nested tori, in turn surrounded by third generation tori, etc; the phase space thus acquires a structure reminiscent of a hierarchical organisation; the carrier of the chaotic solutions is referred to as a *Fat Fractal* (Umberger and Farmer 1985). The time behaviour of stellar oscillations, and the attendant phase space structure, has been studied in the context of various simple ODE models (cf. Takeuti 1989; Perdang 1991) which demonstrate the existence of fractal attractors of dimensions around 2. Adiabatic stellar oscillations exhibit fat fractals (cf. Perdang and Blacher 1982, 1984) similar to those originally isolated by Hénon and Heiles (1964) in celestial mechanics. Instead of a time evolution, we may be directly interested in the ***pattern of the field lines*** defined by an ODE (Eq. 3.12 with the independent variable $t \to \lambda$ an arbitrary variable); for instance, $\mathbf{F}$ may represent a stationary, or a nonstationary velocity field $\mathbf{v}(t,\mathbf{r})$ slowly varying with time t regarded as a parameter, in the 3-dimensional configuration space. The field lines then converge or spiral around specific geometric configurations, namely isolated points, closed curves, toroidal surfaces or fractal surfaces (the counterparts of attractors of the ODE when the parameter λ is intepreted as time). Since a generic function $\mathbf{F}(\mathbf{e};\mathbf{Y})$ produces strange attractors for some parameter ranges, generic field lines likewise trace out fractal structures in the $\mathbf{Y}$-space (or fat fractals in the case of conservative fields). Matter channelled by the field lines of a generic magnetic field, or of a rotation field, will trace out a regular pattern in regions where the field is carried by tori, and distribute irregularly where the carrier of the field is a fractal or (a fat fractal).

The PDE's encountered in an astrophysical context (fluid dynamics, stellar oscillations) are always studied via discretised schemes which reduce the mathematical problem to ODE's (or IM's). As a result, the fact that for instance the density contour lines obtained in numerical simulations of 2D supernova explosions (Müller et al. 1990) are remarkably similar to the fractal displacement fronts encountered in viscous fingering (Feder 1989) merely reflects the property that the spatial difference scheme describing the explosion is algebraically similar to the CA model of viscous fingering. Rigorous reductions of PDE's to ODE's are possible only for PDE's invariant under certain Lie groups (cf. Ovsiannikov 1982, Olver 1986). As an instance, if a PDE in a single space variable x and in time t is invariant under space and time translations, then Morgan's (1952) theorem guarantees that the variable $y = x + Vt$, (V a free parameter) reduces the PDE to an ODE in the independent variable y. Direct studies of fractal sets connected with astrophysical PDE's have been performed by this reduction scheme: Elphick et al. (1990) have studied a 1D heat transport equation as a model for interstellar gas clouds; the energy source function $\epsilon(T, P)$ is chosen such that the equation $\epsilon(T, P) = 0$ has 3 temperature branches $T_1(P) < T_2(P) < T_3(P)$ in the physically relevant pressure range, the intermediate branch T_2 being unstable in a homogeneous medium. The reduced ODE is then shown to describe propagating fronts separating the cold and hot gas components (i.e. clouds) which progressively merge and annihilate. If a forcing term is included in the heat equation, then a chaotic state is generated defining a statistically stationary cloud distribution. Attractor dimensions have not been estimated.

We finally observe that ODE's and PDE's impose on their solutions a strong mathematical constraint of smoothness in time and in space-time (almost everywhere), a requirement which is absent from CA's. Smoothness of the solution tends to suppress the easy growth of (pre)fractal behaviour which becomes visible only in the limit of very large times (and over very large spatial extensions). In CA's prefractal behaviour is already evident after a small number of timesteps.

4. Fractal Candidates in Astronomy

General theoretical arguments are indicative that the following astronomical structures display fractal geometries.

Solar system.– If the surface structures of the terrestrial planets, of the moon and of the major satellites of Jupiter and Saturn are interpreted as instantaneous sections of generic momentum density fields, then these features are expected to be fractals (characterised by a single global fractal mass dimension); with the exception of the earth's surface morphology (cf. Goodchild 1982, Clarke 1986), systematic measurements of the fractal dimensions of planetary reliefs are lacking. For a similar reason (generic momentum density field lines in space), the turbulent atmospheric structures of the Jovian planets should be fractals. The distribution of the diameters of impact craters is presumably directly related to an approximate hierarchical arrangement of the craters (cf. our remarks under 2; cf. also Woronow 1981), and can be described by a global dimension. If Hénon's (1988, 1989) ring models are adequate, the rings of Saturn (cf. Cuzzi et al. 1984; Collins et al. 1984), Jupiter and Uranus (Burns et al. 1984) have the local topology of a direct product of an interval times a Cantor set. Different physical mechanisms are currently thought to shape different parts of the rings: the gaps in Saturn's C ring are identified as resonances; the sharp outer edges of the B and A rings coincide with a 2 : 1 and a 7 : 6 resonance of Mimas and Janus respectively; the 'thousand ringlet structure' is associated with viscous and gravitational effects. Similarly, in the asteroid belt, the Kirkwood gap is related to a resonance of Jupiter (Wisdom 1982; cf. also Franklin et al. 1984). [11] The elements in the solar system (as well as in stars and in interstellar matter) are probably carried by fractal supports, and so are, presumably, the various molecular species isolated in meteorites, in molecular clouds, in the high density and temperature regions near young stars, and in the envelopes of cool stars (cf. the the instantaneous field lines of the partial momentum density fields $\rho_i \mathbf{v}$, ρ_i matter density of species i). [12] Monomolecular or bimolecular reactions taking place in reactors which are not 'well stirred' may generate evolutions of the abundances of the form $e^{-(t/\tau)^a}$ (Kohlrausch-Williams-Watt law) and $(\frac{t}{\tau})^b$ respectively, with $a, b \neq 1$ (Zumofen et al. 1988); such time patterns suggest again a fractal space distribution of the reactants. It remains to be seen whether such spatial fractal supports can develop in the thermonuclear reactions of stellar energy generation. The application of image analysis techniques, quantifying by fractal dimensions the substructures discernible in comets (cf. the catalogue by Rahe et al. 1969), may help clarify the formation process of comets, and of the solar system in general. For instance, the hairy structure of tail streamers in type I tails (cf. Halley's comet) can be enhanced by successive dilations and erosions, so that a global tree dimension and a filament dimension (> 1) can be evaluated; the knots ('fractal dust'), isolated by the operation of erosion eliminating the filaments, can again be described by a fractal dimension (< 1). More importantly, the spongy structure of cometary dust (diameter $\approx 1\mu$m) (Greenberg and Hage 1990), is reminiscent of the fractal geometry of laboratory aggregation clusters (cf. Brandt and Chapman 1981, and Nuth and Donn 1982). The growth of the latter is numerically simulated by stochastic 2-state CA's (Meakin 1986, Witten 1986, Matsushita 1989): in the DLA model, one starts with an initial cluster of m (≥ 1) excited cells; the following steps are then iterated a large number times. A randomly chosen cell at a large distance from the cluster is excited; the excitation propagates in a random walk until it enters a neighbour cell of the cluster; a new cluster is then formed by the attachment of the excited neighbour cell to the previous cluster. In a 3D CA the resulting cluster has a fractal mass dimension ≈ 2.5 (Meakin 1986). According to current ideas, the formation of planets and comets goes through a cascade of aggregation processes (Spinrad 1987): dust grains (frozen H_2O, CO_2, N_2,...) aggregate on a solid core (metal, graphite, silicates and carbon compounds), to form clusters (cf. the DLA model); the latter in turn conglomerate into planetesimals (diameter ≈ 1 km), which in part aggre-

[11] A nonlinear differential equation with a forcing term of frequency Ω shows a periodic behaviour of frequency ω, such that the plot of the ratio ω/Ω against the excitation frequency Ω is a fractal curve (*devil's staircase*, cf. Bergé et al. 1986).

[12] In geophysics there are indeed indications that the Na- and K-content (Wales, Australia) of the soil, and the distribution of iron ore (France) are fractal, the dimensions of the carriers being 1.7-1.9, 1.6 and 1.4-1.9 respectively (Burrough 1981, 1989); cf. also Paliwal et al.'s (1986) estimate for Pb+Zn (India).

gate, to become planets, or remain free ('Oort cloud'). When ejected from the Oort cloud, and when approaching the sun, a planetesimal turns into a comet as a result of the evaporation of the frozen dust grains; the cometary nucleus should acquire the structure of an aerogel (dimension < 2.5). The patterns on the sun's surface (distribution of sunspots, individual sunspot geometry, etc) can be captured by fractal dimensions (island chain dimension, filament dimension) (cf. the theoretical argument in Thieberger et al. 1990).

Stellar distributions, molecular clouds.– The approximation of the solar neighbourhood *Initial Mass Function* $\Phi(m)$ by Salpeter's power law, $\sim m^{-2.35}$ (cf. Tinsley 1980) is suggestive of a hierarchical condensation process (cf. the discussion in Section 2); it is also indicative that 3D CA percolation models, in which a cluster of m cells simulates a star of mass m, could capture the mechanism of protostar formation. Regarding the structure of molecular clouds, our general comment on the atmospheric features, or on the spatial distribution of the elements or molecular species applies, suggesting that these structures are fractals. The genericity argument does not isolate any particular physical mechanism; the fact that the variety of molecular clouds (cf. the Orion and Taurus clouds; Genzel and Stutzki 1989, Shu et al. 1987) have the same type of fragmented shapes as terrestrial clouds, is suggestive that a similar mechanism (hydrodynamic turbulence, diffusion) is operative in both cases. If, as assumed in Perdang (1990b), a diffusion process is responsible for the shape (CO diffusing in a HI matrix in the Orion clouds), then the projection of the diffusion front onto a plane has a fractal mass dimension of 1.34. Alternatively, turbulence, as proposed by Hentschel and Procaccia (1984) for terrestrial clouds, leads to slightly higher values (1.37-1.41). Observational estimates of the fractal dimension by Bazell and Désert (1988) and Blacher and Perdang (1990; ≈ 1.37) are compatible with the values found for terrestrial clouds (≈ 1.35, Lovejoy 1982; Rys and Waldvogel 1986); they remain too crude to distinguish between the two competing mechanisms. CA models for the evolution of the stellar components in a galaxy suggest that the spatial distribution of the red giants and of the protostars may be fractal (Section 3). An observational measurement of these dimensions provides constraints on the mechanisms of formation and evolution of the different stellar components.

Stellar variability.– The traditional use of Fourier series with a small number of expansion coefficients to represent light or velocity curves of variable stars is inadequate to describe chaotic time series $S(t)$. In the standard view the observed time series are reproducible by (fully deterministic) ODE's, so that the time behaviour $S(t)$ itself is not a fractal; the compressed series (cf. above), or alternatively the stroboscopic time series obtained by sampling the chaotic signal $S(t)$ with a timestep of the order of the cycle length determines a fractal curve in the $S(t_i) - t_i$ plane, (t_i, sampling times); Barnsley's (1988) IFS formalism might be used as an interpolation and extrapolation scheme for the fractal curve.

Supernova remnants.– The analogy between viscous fingers in porous media on the one hand, and the density profiles of hydrodynamic simulations of *incipient* supernova explosions on the other is suggestive that CA models capable of describing the *late* evolutionary stages of the explosion can be designed; therefore the spatial matter distribution in the later phases of the explosion is expected to be fractal. Whether the variety of shapes of the observed SNRs just mirrors age differences of these structures (cf. the compact and young Crab nebula, 10^3 y), or whether inherent physical factors are also determinant, might be decided by a comparison of observational and theoretical fractal dimensions, which so far are lacking (cf. however a preliminary measure of a filament dimension of the Vela SNR ≈ 1.20; Blacher and Perdang 1990).

Galactic morphology.– CA models describing stellar evolution in galaxies, beginning with the simplest 2-state model of star formation by Gerola and Seiden (1978), are suggestive that the galactic morphologies are fractals. A satisfactory classification scheme should include therefore a collection of fractal dimensions. Modern classification codes (cf. Thonnat 1989, Buta 1989), just like the original Hubble classification, rely on a few conventional geometrical parameters only. [13] The information of this code is insufficient to recover the galactic pattern. A complementary knowledge of just a few fractal dimensions, however, should supply the missing information for galaxy simulations (cf. the simulation of clusters in Perdang 1990c).

5. Concluding Remarks

Until recently astronomical objects and patterns, ranging from the shapes of planets to the structure of clusters of galaxies, or from the geometry of lightcurves of variable stars to X-ray signals, were captured by a few archetypal geometrical structures (manifolds). Those patterns which defied a representation in terms of the prototypes were viewed as marginal, and either completely ignored by theorists (irregular variability of stars), or unduly squeezed into conventional shapes (approximation of galactic arms by circular cylinders, etc). Fractal geometry now supplies us not only with a better adapted class of reference shapes to approximate the observed complex spatial patterns of turbulent flows, of sunspots or of irregular galaxies and nebulae. It raises a new theoretical problem as well: the geometry of an astronomical configuration was largely considered a subordinate question; the spatial symmetry was adopted at the outset (cf. spherical symmetry of stars; ellipsoidal self-gravitating figures, etc). With the awareness of the varieties of fractal geometries, the selection of one specific fractal pattern becomes in turn a major theoretical problem. We have emphasised in this paper that CA models naturally account for the ubiquity of fractal shapes in space-time. By introducing

[13] In the hierarchical code a galaxy is identified by a chain of 4 characters, $spqT$; s specifies the shape of the galaxy; $s = E$ (elliptical), L (lenticular), S (spiral); the letters p, q are 3-valued logical variables specifying the presence, 'undecidability', or absence of a geometric feature: p refers to bars; $p = B$ (clearly defined bar), X (ill-defined bar), A (absence of a bar); q refers to the ring structure; $q = R$ (clearly defined ring), T (ill-defined ring), S (no ring); the character T is an integer morphological 'type' number, in the range -5 to $+10$ measuring the departure from pure ellipticity ($T = 10$: irregular galaxy).

a minimum of mathematical constraints (which may oppose the formation of fractals, cf. the spatial smoothness requirement), the CA appears at the moment as the most promising framework for numerically modelling fractal structures.

The symbolic notation we have presented makes it evident that the fractal shapes discussed in the literature and simulated numerically so far form just the simplest of a collection containing representatives of arbitrary complexity. It is obvious, however, that CA's are able to generate high complexity fractals, provided only that the neighbourhood is chosen large enough.

This work was supported, in part, by a Royal Society-FNRS European Exchange Grant, 1990.

References

Alexander, S., Orbach, R. (1982): *J. Phys.* **43**, L625.

Alexander, S., Entin-Wohlman, O., Orbach, R. (1986): *Phys. Rev.* **B34**, 2726.

Arnéodo, A., Argoul, F., Elezgaray, J., Grasseau, G. (1988): in *Proc. Conf. on Nonlinear Dynamics*, Ed. G.G. Turchetti.

Arnold, V.I. (1976): *Méthodes mathématiques de la mécanique classique* (Mir, Moscow).

Atmanspacher, H., Scheingraber, H., Voges, W. (1988): *Phys. Rev.* **A37**, 1314.

Auvergne, M., Baglin, A. (1986): *Astron. Astrophys.* **168**, 188.

Avnir, D. (Ed.) (1989): *The Fractal Approach to Heterogeneous Chemistry: Surfaces, Colloids, Polymers* (Wiley, New York).

Bahcall, N.A., Soneira, R.A. (1983): *Astrophys. J.* **270**, 20.

Bahcall, N.A. (1988): *Ann. Rev. Astron. Astrophys.* **26**, 631.

Bak, P., Chen, K. (1989): *Physica* **D38**, 5.

Baltes, H.P., Hilf, E.R. (1978): *Spectra of Finite Systems* (Bibliographisches Institut, Mannheim).

Barenblatt, G.I. (1979): *Similarity, Self-similarity and Intermediate Asymptotics* (Consultants Bureau, New York).

Barnsley, M.F. (1986): *Constructive Approximation* **2**, 303.

Barnsley, M.F. (1988a): *Fractals Everywhere* (Academic Press, Boston).

Barnsley, M.F. (1988b): in *The Science of Fractal Images* Ed. H.O. Peitgen, D. Saupe (Springer-Verlag, New York), p. 219.

Bazell, D., Désert, F.X. (1988): *Astrophys. J.* **333**, 353.

Beck, C. (1990): *Physica* **D41**, 67.

Bergé, P., Pomeau, Y., Vidal, C. (1986): *Order Within Chaos* (Wiley, New York).

Blacher, S., Perdang, J. (1981): *Physica* **D3**, 512.

Blacher, S., Perdang, J. (1990): in *Fractals in Astronomy*, Ed. A. Heck, *Vistas in Astron.* **33**, 393.

Brandt, J.C., Chapman, R.D. (1981): *Introduction to Comets* (Cambridge Univ. Press, Cambridge).

Briggs, K. (1990): *Phys. Letters* **A151**, 27.

Burns, J.A., Showalter, M.R., Morfill, G.E. (1984): in *Planetary Rings*, Eds. R. Greenberg, A. Brahic (Univ. Arizona Press, Tucson), p. 200.

Burrough, P.A. (1981): *Nature* **294**, 240.

Burrough, P.A. (1989): in *The Fractal Approach to Heterogeneous Chemistry*, Ed. D. Avnir (Wiley, New York), p. 383.

Buta, R. (1989): in *The World of Galaxies*, Eds. H.G. Corwin, L. Bottinelli (Springer-Verlag, New York), p. 29.

Calzetti, D., Einasto, J., Giavalisco, M., Ruffini, R., Saar, E. (1987): *Astrophys. Sp. Sc.* **137**, 101.

Calzetti, D., Giavalisco, M., Ruffini, R. (1988): *Astron. Astrophys.* **198**, 1.

Cannizzo, J., Goodings, D. (1988): *Astrophys. J. Letters* **334**, L31

Cannizzo, J., Goodings, D., Mattei, J.A. (1990): *Astrophys. J.* **357**, 235.

Casdagli, M. (1989): *Physica* **D35**, 335.

Cates, M.E., Deutsch, J.M. (1987): *Phys. Rev.* **A25**, 4907.

Charlier, C.V.L. (1922): *Medd. Lund Observ.* **98**.

Clarke, K.C. (1986): *Comput. Geosciences* **12**, 713.

Coleman, P.H., Pietronero, L., Sanders, L.R.H. (1988): *Astron. Astrophys.* **200**, 32.

Collins, S.A., Diner, J., Garneau, G.W., Lane, A.L., Synnott, S.P., Terrile, R.J., Holberg, J.B., Smith, B., Tyler, G.L. (1984): in *Planetary Rings*, Eds. R. Greenberg, A. Brahic (Univ. Arizona Press, Tucson), p. 737.

Courtens, E., Vacher, R., Stoll, E. (1989): *Physica* **D38**, 41.

Cuzzi, J.N., Lissauer, J.J., Esposito, L.W., Holberg, J.B., Marouf, E.A., Tyler, G.L., Boischot, A. (1984): in *Planetary Rings*, Eds. R. Greenberg, A. Brahic (Univ. Arizona Press, Tucson), p. 73.

Davis, M., Peebles, P.J.E. (1983): *Astrophys. J.* **267**, 465.

Devaney, R. (1986): *An Introduction to Chaotic Dynamical Systems* (Addison-Wesley, New York).

de Vaucouleurs G. (1970): *Science* **167**, 1203.

de Vaucouleurs G. (1971): *Publ. Astron. Soc. Pacific* **83**, 113.

de Vaucouleurs G. (1974): in *The Formation and Dynamics of the Galaxies* (D. Reidel Publ. Co. Dordrecht), p. 1.

Essam, J.W. (1980): *Rep. Progr. Phys.* **43**, 833.

Falconer, K.J. (1985): *The Geometry of Fractal Sets* (Cambridge Univ. Press, Cambridge).

Favard, J. (1950): *Espace et Dimension* (Albin Michel, Paris).

Feder, J. (1989): *Fractals* (Plenum Press, New York).

Feigenbaum, M.J. (1979): *J. Stat. Phys.* **21**, 669.

Fourcade, B., Breton, P., Tremblay, A.M.S. (1987): *Phys. Rev.* **B36**, 8925.

Fourcade, B., Tremblay, A.M.S. (1987): *Phys. Rev.* **A36**, 2352.

Franklin, F., Lecar, M., Wiesel, W. (1984): in *Planetary Rings*, Eds. R. Greenberg, A. Brahic (Univ. Arizona Press, Tucson), p. 562.

Genzel, R., Stutzki, J. (1989): *Ann. Rev. Astron. Astrophys.* **27**, 41.

Gerola, H., Seiden, P.E. (1978): *Astrophys. J.* **223**, 129.

Goldreich, P., Tremaine, S. (1982): *Ann. Rev. Astron. Astrophys.* **20**, 249.

Goodchild, M.F. (1982): in *Proc. Modeling and Simulation Conf.* **3**, 1133.

Grassberger, P. (1985): in *Chaos in Astrophysics*, Eds. J.R. Buchler, J.M. Perdang, E.A. Spiegel, NATO ASI **C 161** (D. Reidel Publ. Co., Dordrecht), p. 193.

Grassberger, P., Procaccia, I. (1983a): *Physica* **D9**, 189.

Grassberger, P., Procaccia, I. (1983b): *Phys. Rev. Letters* **50**, 346.

Greenberg, J.M., Hage, J.I. (1990): *Astrophys. J.* **361**, 260.

Halsey, T.C., Jensen, M.H., Kadanoff, L.P., Procaccia, I., Shraiman, B.I. (1986): *Phys. Rev.* **A33**, 1141.

Hayashi, C., Hoshi, R., Sugimoto, D. (1962): *Progr. Theor. Phys. Suppl.* **22**.

Hentschel, H.G.E., Procaccia, I. (1983): *Physica* **D8**, 435.

Hentschel, H.G.E., Procaccia, I. (1984): *Phys. Rev.* **A29**, 1461.
Hénon, M. (1989): *La Recherche* **20**, 490.
Hénon, M. (1988): *Physica* **D33**, 132.
Hénon, H., Heiles, C. (1964): *Astron. J.* **68**, 73.
Herrmann, H.J. (1986): in *On Growth and Form*, Eds. H.E. Stanley, N. Ostrowsky (Martinus Nijhoof, Dordrecht), p. 3.
Ikeuchi, S., Yoshioka, S. (1989): *Comments Astrophys.* **13**, 117.
Jaynes, E.T. (1957): *Phys. Rev.* **106**, 620.
Jones, B.F., Walker, M.F. (1988): *Astron. J.* **95**, 1755.
Jones, B.J.T., Martinez, V.J., Saar, E., Einasto, J. (1988): *Astrophys. J.* **332**, L1.
Kac, M. (1966): *Amer. Math. Monthly* **73S**, 1.
Kolmogorov, A.N. (1965): *Problemy peredachi informatsii* **1**, 3.
Kolláth, Z. (1990): *Monthly Not. Roy. Astron. Soc.*, in press.
Kovács, G., Buchler, J.R. (1988): *Astrophys. J.* **334**, 971.
Ledoux, P., Walraven, P. (1958): *Handbuch der Physik* **51**, 353.
Lejeune, A., Perdang, J. (1990): in *Proc. European Conf. Computational Physics.*
Lejeune, A., Perdang, J. (1991): *CA Simulations of Stellar Distributions*, in press.
Liu, S.H. (1986): *Solid State Phys.* **39**, 207.
Lovejoy, S. (1982): *Science* **216**, 185.
Mandelbrot, B.B. (1975): *Les objets fractals* (Flammarion, Paris).
Mandelbrot, B.B. (1977): *Fractals: Form, Chance, Dimension* (Freeman, San Francisco).
Mandelbrot, B.B. (1982): *The Fractal Geometry of Nature* (Freeman, San Francisco).
Mandelbrot, B.B., Given, J.A. (1984): *Phys. Rev. Letters* **52**, 1853.
Mandelbrot, B.B. (1989): in *The Fractal Approach to Heterogeneous Chemistry*, Ed. D. Avnir (Wiley, New York), p. 45.
Manneville, P., Boccara, N., Vichniac, G.Y., Bidaux, R. (Eds.) (1989): *Cellular Automata and Modeling of Complex Physical Systems*, *Springer Proc. Phys.* **46**.
Martin, O., Odlyzko, A., Wolfram, S. (1984): *Comm. Math. Phys.* **93**, 219.
Martinez, V.J., Jones, B.J.T. (1990): *Monthly Not. Roy. Astron. Soc.* **242**, 517.
Massewitsch, A.G., Parenago, P.P. (1951): *Abh. sowjet. Astronomie* **2**, 27.
Matsushita, M. (1989): in *The Fractal Approach to Heterogeneous Chemistry*, Ed. D. Avnir (Wiley, New York), p. 161.
May, R.B. (1976): *Nature* **261**, 456.
May, R.B. (1983): in *Chaotic Behaviour of Deterministic Systems*, NATO ASI (North Holland, Amsterdam), p. 513.
Mayer-Kress, G. (1987): in *Springer Series in Synergetics* **32**, 246.
Meakin, P. (1986): in *On Growth and Form*, Eds. H.E. Stanley, N. Ostrowsky (Martinus Nijhoof, Dordrecht) p. 111.
Mera, C. (1987): *Chaotic Dynamics* (World Scientific, Singapore).
Morgan, A.J.A. (1952): *Quart. J. Math.* **3**, 250.
Müller, E., Fryxell, B., Arnett, D. (1990): in *Chemical and Dynamical Evolution of Galaxies*, Eds. F. Ferrini, F. Matteucci, J. Franco (Giardini, Pisa).
Newman, W.I., Wasserman, I. (1990): *Astrophys. J.* **354**, 411.
Nottale, L. (1989): *Intern. J. Modern Physics* **A4**, 5047.
Nuth, J.A., Donn, B. (1982): *J. Chem. Phys.* **77**, 2639.
Olver, P.J. (1986): *Application of Lie Groups to Differential Equations* (Springer-Verlag, New York).
O'Shaughnessy, B., Procaccia, I. (1985): *Phys. Rev. Letters* **54**, 455.

Ovsiannikov, L.V. (1982): *Group Analysis of Differential Equations* (Academic Press, New York).
Ostriker, J.P., Cowie, L.L. (1981): *Astrophys. J. Letters* **243**, L127.
Packard, N.H., Wolfram, S. (1985): *J. Stat. Phys.* **38**, 5/6.
Paliwal, H.V., Bhatnagar, S.N., Haldar, S.K. (1986): *J. Math. Geology* **18**, 539.
Parenago, P.P., Massewitsch, A.G. (1951): *Abh. sowjet. Astronomie* **2**, 27.
Pawelzik, K., Schuster, H.G. (1987): *Phys. Rev.* **A35**, 481.
Peebles, P.J.E. (1980): *The Large-Scale Structure of the Universe* (Princeton Univ. Press, Princeton).
Peebles, P.J.E. (1989): *Physica* **D38**, 273.
Peitgen, H.O., Saupe, D. (1988): *The Science of Fractal Images* (Springer-Verlag, New York).
Peitgen, H.O., Richter, P.H. (1986): *The Beauty of Fractals* (Springer-Verlag, New York).
Pellegrini, P.S., Willmer, C.N.A., Da Costa, L.N., Santiago, B.X. (1990): *Astrophys. J.* **350**, 95.
Perdang, J. (1981): *Astrophys. Sp. Sc.* **74**, 149.
Perdang, J. (1982): *Astrophys. Sp. Sc.* **83**, 311.
Perdang, J. (1988): *Fractals and Chaos*, Lecture Notes, Maîtrise Astrophys. Géophys. (Univ. Liège).
Perdang, J. (1989): in *The Numerical Modelling of Stellar Pulsations*, NATO ASI **C 302**, 333.
Perdang, J. (1990a): *Ann. New York Acad. Sc.*, in press.
Perdang, J. (1990b): in *Fractals in Astronomy*, Ed. A. Heck, *Vistas Astron.* **33**, 249.
Perdang, J. (1990c): in *Fractals in Astronomy*, Ed. A. Heck, *Vistas Astron.* **33**, 371.
Perdang, J. (1991): in *Proc. ESO Workshop Rapid Variability of OB-Stars: Nature and Diagnostic Value* (ESO, Garching bei München).
Perdang, J., Blacher, S. (1982): *Astron. Astrophys.* **112**, 35.
Perdang, J., Blacher, S. (1984): *Astron. Astrophys.* **136**, 263.
Petit, J.M., Hénon, M. (1986): *Icarus* **66**, 536.
Pierre, M. (1990): *Astron. Astrophys.* **229**, 7.
Pietronero, L., Tossati, E. (1986): *Fractals in Physics* (Elsevier, Amsterdam).
Prasad, C., Meneveau, C., Sreenivasan, K.R. (1988): *Phys. Rev. Letters* **61**, 74.
Rényi, A. (1970): *Probability Theory* (North-Holland, Amsterdam).
Rahe, J., Donn, B., Wurm, K. (1969): *Atlas of Cometary Forms*, NASA-SP **198**.
Rammal, R., Toulouse, G. (1983): *J. Phys. Lettres* **44**, L13.
Rapaport, D.C. (1985): *J. Phys.* **A18**, L175.
Rasband, S.N. (1990): *Chaotic Dynamics of Nonlinear Systems* (Wiley, New York).
Rood, H.J. (1988): *Ann. Rev. Astron. Astrophys.* **26**, 245.
Rys, F.S., Waldvogel, A. (1986): *Phys. Rev. Letters* **56**, 784.
Schlögl, F. (1980): *Phys. Rep.* **62**, 267.
Sapoval, B. (1989): *Physica* **D38**, 296.
Sapoval, B., Rosso, M., Gouyet, J.F. (1985): *J. Phys. Lettres* **46**, L149.
Sapoval, B., Rosso, M., Gouyet, J.F. (1989): in *The Fractal Approach to Heterogeneous Chemistry*, Ed. D. Avnir (Wiley, New York), p. 227.
Saupe, D. (1988): in *The Science of Fractal Images*, Eds. H.O. Peitgen, D. Saupe (Springer-Verlag, New York), p. 71.
Schulman, L.S., Seiden, P.E. (1982): *J. Stat. Phys.* **27**, 83.
Schulman, L.S., Seiden, P.E. (1983): in *Percolation Structures and Processes*, Eds. G. Deutscher, R. Zallen, J. Adler (Adam Hilger, Bristol), p. 251.

Schulman, L.S., Seiden, P.E. (1986): *Astrophys. J.* **311**, 1.
Seiden, P.E., Gerola, H. (1982): *Fundamentals Cosmic Phys.* **7**, 241.
Serra, J. (1988): *Image Analysis and Mathematical Morphology* Vol. **1** (Academic Press, London).
Shu, F.H., Adams, F.C., Lizano, S. (1987): *Ann. Rev. Astron. Astrophys.* **25**, 23.
Slezak, E., Bijaoui, A., Mars, G. (1990): *Astron. Astrophys.* **227**, 301.
Spinrad, M. (1987): *Ann. Rev. Astron. Astrophys.* **25**, 231.
Stauffer, D. (1985): *Introduction to Percolation Theory* (Taylor and Francis, London).
Stauffer, D. (1986): in *On Growth and Form*, Eds. H.E. Stanley, N. Ostrowsky (Martinus Nijhoff, Dordrecht), p. 79.
Stone, E.C., Miner, E.D. (1981): *Science* **212**, 159.
Takens, F. (1981): in *Dynamical Systems and Turbulence*, *Lecture Notes Math.* **898** (Springer-Verlag, Berlin) p. 366.
Takeuti, M. (1989): in *The Numerical Modelling of Stellar Pulsations*, NATO ASI **C 302**, 121.
Temperley, H.N.V. (1981): *Graph Theory and Applications* (Wiley, New York).
Thieberger, R., Spiegel, E.A., Smith, L.H. (1990): in *The Ubiquity of Chaos* (Amer. Ass. Adv. Sc., Washingthon), p. 197.
Thonnat, M. (1989): in *The World of Galaxies*, Eds. H.G. Corwin, L. Bottinelli (Springer-Verlag, New York), p. 53.
Todd, J.A. (1947): *Projective and Analytical Geometry* (Pitman, London).
Tinsley, B.M. (1980): *Fundamentals Cosmic Phys.* **5**, 287.
Umberger, D.K., Farmer, J.D. (1985): *Phys. Rev. Letters* bf 55, 661.
Vicsek, T., Szalay, A.S. (1987): *Phys. Rev. Letters* **58**, 2818.
Voss, R.F. (1985): in *Fundamental Algorithms in Computer Graphics*, Ed. R.A. Earnshaw (Springer-Verlag, Berlin), p. 805.
Voss, R.F. (1988): in *The Science of Fractal Images*, Eds. H.O. Peitgen, D. Saupe (Springer-Verlag, New York), p. 21.
Wheeler, J.A. (1962): *Geometrodynamics* (Academic Press, New York).
Weyl, H. (1911): *Nachr. königl. Ges. Wiss. Göttingen*, Math.-phys. Kl. **110**
Wisdom, J. (1982): *Astron. J.* **87**, 577.
Witten, T.A. (1986): in *On Growth and Form*, Eds. H.E. Stanley, N. Ostrowsky (Martinus Nijhoff, Dordrecht) p. 54.
Wolfram, S. (1983): *Rev. Modern Phys.* **55**, 601.
Wolfram, S. (1984): *Physica* **D10**, 1.
Wolfram, S. (1985): in *Dynamical Systems and Cellular Automata*, Eds. J. Demongeot, E. Golès, M. Tchuente (Academic Press, London), p. 253.
Woronow, A. (1981): *Math. Geology* **13**, 201.
Yoshioka, S., Ikeuchi, S. (1989): *Astrophys. J.* **341**, 16.
Zeldovich, Ya.B., Ruzmaikin, A.A., Sokoloff, D.D. (1990): *The Almighty Chance*, Lecture Notes Phys. **20** (World Scientific, Singapore).
Zumofen, G., Blumen, A., Klafter, J. (1988): in *Fractals, Quasicrystals, Knots and Algebraic Quantum Mechanics*, NATO ASI C **235**, 1.

Pulsating Stars and Fractals

Marie-Jo Goupil, Michel Auvergne, Thierry Serre

Observatoire de Paris, U.R.A. C.N.R.S. 335, 5 Place Jules Janssen,
F–92195 Meudon, France

Abstract: Fractals are encountered in stellar pulsation theory as attractors, resulting from chaotic dynamics of pulsating stars. We review observational and theoretical investigations of chaotic stellar pulsations with particular emphasis on the physical information they provide about stellar interiors.

I. Introduction

Pulsating stars owe their names to the temporal variations of their luminosities and radial velocities. Since these variations originate from internal physical mechanisms, the study of stellar pulsations, through a tight confrontation between observations and theory, efficiently probes the internal structure and evolution of these stars. For a periodic pulsating star such as a classical Cepheid, the period of pulsation is found of the order of the free fall time $\tau \sim (G\rho)^{-1/2}$ (Cox and Giuli 1968) and therefore provides an estimation of its average density ρ, (G gravitational constant). When the oscillation displays two simultaneous periods, a determination of the stellar mass is available, independently of the evolution theory (Petersen 1973). With recent results from multisite campaigns, asteroseismology of white dwarfs with hundreds of observed frequencies starts to be fruitful (Winget 1990). Thousands of pulsation frequencies have been detected in the solar case and helioseismology has already brought up valuable informations about the depth of the convective zone and the internal rotation of the sun (Christensen-Dalsgaard 1988; Provost and Berthomieu 1990).

In Sect. II, we review our current knowledge of stellar radial pulsations. Our theoretical understanding comes from intensive investigations of the partial differential equations of hydrodynamics and heat transfer, which model the stellar envelope. When the oscillation amplitude is small enough, a linearization of the hydrodynamical system has led to identify the destabilizing mechanisms and the type of pulsating modes for each group of pulsating stars (radial, non radial modes) (II-1). In this process, it has been enlightening to study simplified linear models in which only the basic physics of the pulsation is retained (II-2). However linearization can account neither for the observed finite amplitudes of oscillation nor for the selection of modes. These are nonlinear processes that must be investigated through the full hydrodynamical system. Following the initial work of Christy (1966), many hydrodynamical models of radially pulsating stars have been computed. They reproduce the main features of the pulsation, indicating that their physical content models stellar envelopes rather correctly (II-3). Saturation and selection mechanisms limit the amplitude increase of an unstable mode, favor a few number of linearly unstable modes and nonlinearly damp the others. This can be studied by means of a semi analytical approach which gives rise to the so called amplitude equations. This small set of ordinary differential equations governs the amplitude evolution of the relevant modes of pulsation (II-4).

For a long time, most of the variable stars have been thought to be more or less periodic, but, with the advent of increasingly accurate observations and extended time series, many stars have been found to be multiperiodic or irregular (for a review, see Perdang 1985). Until recently this irregular behavior was either dismissed as noise or discarded because of a lack of knowledge of a plausible physical cause. In the last years however, a great deal of attention has been focused on irregular stellar pulsators. Indeed results of the dynamical system theory tell us that simple (few degrees of freedom) dissipative dynamical systems can exhibit very complex, irregular temporal variations (Cvitanović et al. 1984; Bergé et al. 1986). This deterministic chaos arises from well known routes to chaos when varying a control parameter of the system and proves to be a quite common phenomenon, observed in various fields of physics. One thus can legitimately wonder whether such chaotic dynamics also exist for the dissipative dynamical systems that are pulsating stars and are responsible for some of the observed irregular behaviors. Hydrodynamical stellar models do indeed display chaotic pulsations (II-3) and some possible physical causes for the existence of stellar chaos have been investigated via simplified nonlinear models (II-5.1). Actually a rich set of regular and chaotic stellar behaviors ought to be expected as shown by the study of a quadratic map stemming from the original system (II-5.2).

Thus chaos can definitely exist in stellar conditions and this raises the question of how to recognize chaotic dynamics from observational data of variable stars i.e. from noisy and time limited series of luminosity or velocity measurements. From the theory of dissipative dynamical systems, chaotic attractors are known to be characterized by two major features: the exponential divergence of trajectories in phase space, which controls the long term unpredictibility of the system, and, in most cases, their fractal structures. Noninteger dimension(s) and (local) scaling invariance at small length scales are specific properties of a fractal object while the Lyapunov exponents measure the exponential divergence. Procedures to estimate these quantities which must be adapted to the difficulties encountered with observational data are surveyed in Sect. III. More details may be found in Auvergne et al. (1990a), Perdang (1990a) and references therein.

Attempts to use these procedures to identify chaos in stellar variations are reported in Sect. IV and conclusions are addressed in Sect. V.

II. Pulsating Stars

The basic hydrodynamical and heat transfer equations that model a spherically symmetric star are, in a lagrangean description, (Cox 1980a):

$$\frac{d^2r}{dt^2} = -4\pi r^2 \frac{\partial P}{\partial m} - \frac{Gm}{r^2} \quad \text{momentum equation} \tag{1a}$$

$$\frac{ds}{dt} = -\frac{1}{T}\frac{\partial L}{\partial m} \quad \text{energy conservation equation} \tag{1b}$$

$$\frac{1}{\rho} = 4\pi r^2 \frac{\partial r}{\partial m} \quad \text{continuity equation} \tag{1c}$$

$$L = -(4\pi r^2)^2 \frac{ac}{3\kappa}\frac{\partial T^4}{\partial m} \quad \text{radiative luminosity} \tag{1d}$$

where m, the mass within a sphere of radius r, and time t are the lagrangean coordinates and r, s, ρ, T, L, κ denote the radius, entropy, density, temperature, luminosity, and opacity respectively. As written here, System (1) only holds for radiative transfer. Nuclear sources of energy, known to play a negligible part in the pulsation process, and convective transport are disregarded. The star is assumed to maintain its spherical symmetry in the course of its pulsations i.e. undergo radial pulsations. For non radial pulsations, see Unno et al. (1989). Constitutive equations i.e. equation of state $P = P(\rho, T)$ and opacity $\kappa = \kappa(\rho, T)$ plus boundary and initial conditions complete the description. The thermodynamical variables are chosen to be s and ρ, hence $T = T(\rho, s)$. Equilibrium (i.e. time independent) quantities, specified with an overbar, verify:

$$-4\pi\bar{r}^2\frac{\partial \bar{P}}{\partial m} = \frac{Gm}{\bar{r}^2} \equiv g; \qquad \frac{\partial \bar{L}}{\partial m} = 0; \qquad \frac{1}{\bar{\rho}} = 4\pi\bar{r}^2\frac{\partial \bar{r}}{\partial m} \tag{2}$$

For weakly nonlinear (i.e. small amplitude) oscillations, one assumes small departures from equilibrium such that:

$$r = \bar{r}\,(1 + \frac{\delta r}{\bar{r}}); \quad |\frac{\delta r}{\bar{r}}| < 1; \qquad s = \bar{s} + \delta s; \quad |\delta s| < \bar{s} \tag{3}$$

A Taylor expansion of the right hand side of Eqs. (1) is then carried out about the equilibrium (2). A quantity such as the density, for instance, becomes:

$$\rho = \bar{\rho}\,(1 + \frac{\delta_1\rho}{\bar{\rho}} + \frac{\delta_2\rho}{\bar{\rho}} + \frac{\delta_3\rho}{\bar{\rho}} + ...) \tag{4}$$

where δ_i means a lagrangean variation of order i in $\delta r/r$ and/or δs. The resulting system, expanded up to third order, is cast into the form:

$$\frac{d\mathbf{z}}{dt} = \mathbf{L}(\mathbf{z}) + \mathbf{N_2}(\mathbf{z}, \mathbf{z}) + \mathbf{N_3}(\mathbf{z}, \mathbf{z}, \mathbf{z}) \tag{5}$$

where $\mathbf{z} = (\delta r/r, v \equiv d(\delta r/r)/dt, \delta s/c_v)$; c_v is the specific heat at constant volume and $\mathbf{L}(\mathbf{z}) \equiv (L_r, L_v, L_s)$ is the usual nonadiabatic linear operator. The vectors

$\mathbf{N_2} \equiv (N_{2r}, N_{2v}, N_{2s})$, $\mathbf{N_3} \equiv (N_{3r}, N_{3v}, N_{3s})$ collect the nonlinear terms of respectively quadratic and cubic orders in $\delta r/r$ and δs. Explicit expressions for $\mathbf{L}, \mathbf{N_2}, \mathbf{N_3}$ are given in App. A.

II.1 Linear Stability Analysis

Small enough amplitude variations are satisfactorily described by the linearized system

$$\frac{d\mathbf{z}}{dt} = \mathbf{L}(\mathbf{z}) \tag{6}$$

Assuming a time dependence: $\mathbf{z} = \xi(m)\mathrm{e}^{\sigma t}$, $\xi \equiv (\xi_r, \xi_v, \xi_s)$, System (6) with linearized boundary conditions gives rise to an eigenvalue problem in which $\xi_j(m)$ is the eigenvector (normalized to unity at the surface $m = M$) associated with the eigenvalue $\sigma_j = i\Omega_j + \kappa_j$, where Ω_j is the pulsation frequency and κ_j the amplitude growth or decay rate of the radial pulsation mode ξ_j. When $\kappa_j > 0$, the pulsation mode ξ_j is said to be vibrationally unstable. One notes that:

$$\sigma\xi_r = \xi_v \tag{7a}$$

$$\sigma\xi_v = L_v(\xi) = -4\pi\bar{r}\frac{d\delta_1 P}{dm} + 4\frac{g}{\bar{r}}\xi_r \tag{7b}$$

$$\sigma\xi_s = L_s(\xi) = -\frac{1}{c_v\bar{T}}\frac{d\delta_1 L}{dm} \tag{7c}$$

In Eq. (7) and from here on, the variation $\delta_i f$ of any quantity f refers either to $\delta_i f(m,t)$ or $\delta_i f(m)$ if no ambiguity occurs.

In the general nonadiabatic case, $\mathbf{L}(\mathbf{z})$ is real but not self-adjoint and the adjoint eigenvectors $\tilde{\xi}_k \equiv (\tilde{\xi}_r, \tilde{\xi}_v, \tilde{\xi}_s)_k$, associated with the eigenfrequencies σ_k, obey the orthogonality relation:

$$\int_M dm\ \tilde{\xi}_k \cdot \xi_j = N_k\ \delta_{kj} \tag{8}$$

One has $\tilde{\xi}_r = \sigma\tilde{\xi}_v$ and the norm is given by

$$N_k = \int_M dm\ (\tilde{\xi}_{rk}\xi_{rk} + \tilde{\xi}_{vk}\xi_{vk} + \tilde{\xi}_{sk}\xi_{sk}) = 2\sigma_k \int_M dm\ \tilde{\xi}_{vk}\xi_{rk} + \int_M dm\ \tilde{\xi}_{sk}\xi_{sk} \tag{9}$$

In the adiabatic approximation, the entropy is constant and the equation of energy simply is $dP/P = \Gamma_1 d\rho/\rho$. System (6) reduces to a second order differential equation in $\delta r/r$. The linear operator is self-adjoint ($\mathbf{L} = \tilde{\mathbf{L}}^{\mathbf{t}}$) and, with appropriate boundary conditions, one deals with a Sturm–Liouville problem for which the eigenfrequencies are purely imaginary $\sigma = \pm i\omega$. The component ξ_r is purely real and $\xi_v = i\omega\xi_r$ (Ledoux and Walraven 1958). The adjoint eigenvector is given by

$$\tilde{\xi}_r = i\omega\tilde{\xi}_v; \quad \tilde{\xi}_v = \bar{r}^2\xi_r; \quad N = 2i\omega J \tag{10}$$

For each mode, one then has:

$$\omega^2 = J^{-1}\int_M dm\ \bar{r}^2\xi_r L_r(\xi_r),$$

where

$$J = \int_M dm\ \bar{r}^2 \xi_r^2 > 0$$

is the moment of inertia. Comparisons between computed and observed frequencies have led to an identification of the classical variables as radial pulsators. Investigations of local destabilizing mechanisms can be carried out by means of a quasiadiabatic approximation, i.e. for $\mu \equiv \bar{L}/\Delta m c_v \bar{T}\omega << 1$, for which entropy and energy terms are evaluated with the adiabatic eigenvectors but nonadiabatic, complex eigenfrequencies $\sigma = \kappa + i\Omega$ are maintained (Ledoux and Walraven 1958; Cox 1980a). The growth rate is thereby obtained as:

$$\kappa = -\frac{1}{2\Omega^2 J}\int_M dm\ (\Gamma_3 - 1)\frac{\delta_1 \rho}{\rho}\ \frac{\partial \delta_1 L}{\partial m} \tag{11}$$

The existence or nonexistence of an efficient destabilizing mechanism determines the sign of κ. It can be inferred from (11) that (κ, γ) instability mechanisms arise from the existence of hydrogen and helium partial ionization zones, their efficiencies depending on the position of these zones in the star (Cox 1980b). This can more simply be seen by studying a simplified model as shown next.

II.2 Baker's One Zone Model

In the classical one zone formalism introduced by Baker (1966), all spatial derivatives are eliminated except for the luminosity. The linearization yields a third order ordinary differential equation:

$$\dddot{x} + A\mu\omega_0\ \ddot{x} + (3\Gamma_1 - 4)\omega_0^2\ \dot{x} + B\mu\omega_0^3\ x = 0 \tag{12}$$

where dots means time derivatives; $x = \delta r/\bar{r}$, $\omega_0^2 = GM/\bar{r}^3$, $A = (\Gamma_3 - 1)(s + 4)$, $B = (\Gamma_3 - 1)(3n - 4s)$; s, n are the opacity exponents when assuming a power law dependence of the absorption coefficient $\kappa = \kappa_0 \rho^n T^{-s}$; Γ_1, Γ_3 are the adiabatic exponents (App. A). The ratio of the luminosity to the internal energy per period, $\mu = \bar{L}/\Delta m c_v \bar{T}\omega_0$, where Δm is the mass of the zone, measures the departure from adiabaticity ($\mu = 0$ in the adiabatic approximation).

Assuming a time dependence of x in exp(σ t), equation (12) becomes a cubic equation for $\sigma = i\Omega + \kappa$. The criterion for a vibrational instability ($\kappa > 0$) depends on A, B, C, hence on the behavior of Γ_1, Γ_3 and the opacity. When $\Gamma_1 < 4/3$, a dynamical instability occurs. Both vibrational and dynamical instabilities are at work in actual stars in some regions of their interiors. The net result depends on the weight of the unstable zones compared to the whole star.

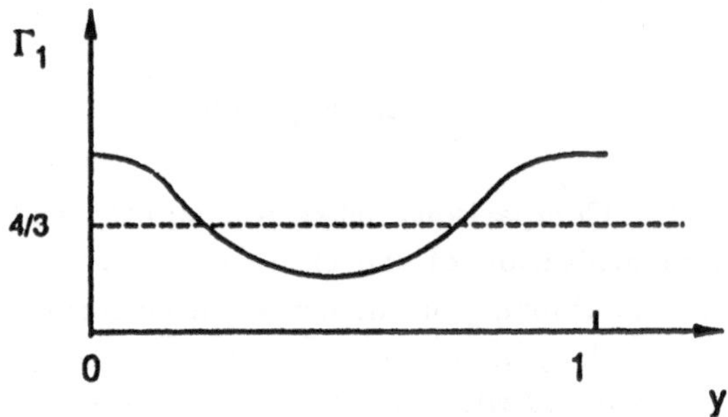

Fig. 1. Variation of Γ_1 as a function of the ionization degree y

II.3 Hydrodynamical Models

Hydrodynamical models of regular radial pulsators such as classical Cepheids and RR Lyrae stars (Stobie 1969a,b,c; King et al. 1973; Stellingwerf 1975; Stothers 1981) have shown that they generally pulsate in the fundamental or the first overtone radial modes. Features such as phase lags and skewness of light and velocity curves agree rather well with the observations. Roles played by ionization zones and dissipative interiors in limiting amplitude increases have been discussed by Christy (1966) and Karp (1975) for instance.

A Fourier decomposition technique, in computing Fourier amplitudes and phases from observed and numerical light and velocity curves, offers a quantitative way of confronting observations and hydrodynamical models (Simon 1988). Such comparisons, carried out for classical variables, have shown that this technique is able to discriminate between fundamental and overtone pulsators (Antonello and Mantegazza 1984) and to detect the presence of resonances such as the period resonance $P_2/P_0 \sim 1/2$ of Bump Cepheids (Simon and Davis 1983; Buchler et al. 1990). Efforts to alleviate some discrepancies brought out by the above comparisons should lead to improve the physical content of hydrocodes (Simon and Aikawa 1986; Simon 1990).

Numerical models of Mira and semi regular stars (Ostlie et al. 1982; Tuchman et al. 1978, 1979) show irregular pulsations. Light curves of pop II Cepheids models have been reported to display irregular alternations of deep and shallow mimima, typical of RV Tauri stars (Christy 1967; Stobie 1969b; Fadeyev and Fokin 1985). That low dimensional chaos can be instrumental in generating such irregularities has been demonstrated with sequences of hydrodynamical models of pop II Cepheids whose transitions to chaos arise either through period doubling cascades or tangent bifurcations (Buchler and Kovács 1987a; Buchler et al. 1987; Aikawa 1987). Return maps for these models are almost one dimensional (Fig. 2) and dimensions of the chaotic attractors are estimated to be smaller than 3 (Kovacs and Buchler 1988).

Insights into the physical origin of these chaotic pulsations can be inferred by means of return maps using energy variables such as the kinetic energy (Takeuti 1987). For instance, $W(t) = (2\Pi N)^{-1} \int_0^t \int_M P(dV/dt')dt'$ is the work integral at

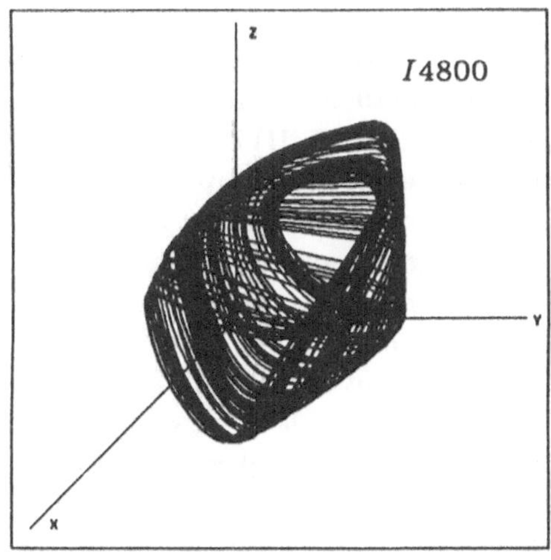

Fig. 2a. A 2D projection of the attractor of a pop. II Cepheid chaotic model (from Kovács and Buchler 1988).

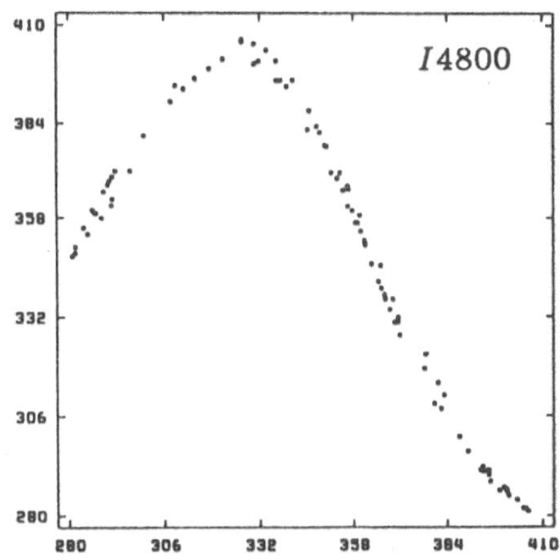

Fig. 2b. An associated return map (from Kovács and Buchler 1988).

time t, Π is the main period, N some normalization factor, $V = 1/\rho$ and 0 an arbitrary time origin. Constructing a return map in plotting $W(t + \Pi)$ against $W(t)$ for every cycle i.e. at $t = n\Pi$, one has $W((n+1)\Pi) = W(n\Pi) + \Delta W_n(\Pi)$ with $\Delta W_n(\Pi) = \int_{n\Pi}^{(n+1)\Pi} \int_M P(dV/dt')dt'$. For a Π periodic limit cycle, no energy is gained from cycle to cycle and $\Delta W_n = 0$, the first return map reduces to one single point, $W((n+1)\Pi) = W(n\Pi)$, on the first bissector. For a period doubled (2Π) limit cycle, the net energy gain is zero only after two cycles of length Π i.e. $\Delta W_n(\Pi) \neq 0$ but $\Delta W_n(2\Pi) = 0$ and $W((n+2)\Pi) = W(n\Pi)$. The first return map is now composed of two points apart from the first bissector while the second return map presents one single point on the bissector. For a chaotic attractor, phases of driving and dissipation can similarly be traced. Phases which gain energy from cycle to cycle ($\Delta W_n > 0$) are signaled out by points above the bissector whereas dissipative phases drop the points below it ($\Delta W_n < 0$).

By inspecting first return maps, using kinetic energy maxima at expanding phases, and studying the linear work integral per zone, Aikawa (1987, 1988) analyzed the origin of the type I intermittency (tangent bifurcation) observed in his models: during the laminar (almost periodic) phases, occuring at amplitudes very close to the amplitude of the 'ghost' limit cycle, every cycle gains very little energy. The amplitude, however slowly increases until an extreme compression of H, He ionization zones sharply enhances the opacity at contracting phases. This strong driving triggers the observed sudden outbursts of large amplitude oscillations. Dissipation by shock waves drags the system back to small amplitude oscillations, a necessary reinjection into the neighborhood of the (ghost) limit cycle (Tresser et al. 1980). The whole process then repeats again but never quite similar to the preceeding one.

In the same way than linear eigenfrequencies measure the (in)stability of a fixed point (equilibrium), the Floquet coefficients quantify the (in)stability of a limit cycle (periodic oscillation). Moskalik and Buchler (1990) carefully studied the evolution of the Floquet coefficients for sequences of (Pop. I, II) Cepheids which undergo period doubling cascades when decreasing the effective temperature. The authors established a correlation between the occurence of particular features of the Floquet coefficients and the presence of half integer resonances. It was found that half integer (5/2; 3/2) resonances between the nonlinear frequencies of the fundamental and the second overtone drive the first period doubling of the cascades in these models. Subsequent period doublings do or do not occur up to full chaos, depending on the strength of the nonadiabaticity of the models as suggested by a simple model of two nonlinearly coupled oscillators (Moskalik and Buchler 1990).

Finally Buchler and coworkers stress that while the exact location of the bifurcations in the Herzsprung Russell diagram is currently model dependent (primarily due to the treatment of shocks by an artificial viscosity), the existence of chaos is robust. Supporting this assertion, not only transitions to chaos survive but their locations in the Herzsprung Russell diagram and the amplitude values get closer to the observations when a phenomelogical treatment of time dependent convection (Glasner and Buchler 1990) and non steady radiative transfer (Fokin 1990) respectively are included in the models.

Hence improvements of numerical and physical contents of the hydrocodes in the last past years have shed new light on stellar pulsations but some stubborn disagreements with the observations subsist. For an up to date review on this subject, see *The Numerical Modelling of Nonlinear Stellar Pulsations, Problems and Prospects*, 1989, **302**, Ed.J.R. Buchler (Kluwer Acad. Publ., Dordrecht).

II.4 Amplitude Equations

Ordinary differential equations generated from the partial differential system (1), the amplitude equations govern the temporal behavior of amplitudes of oscillation. Their derivations rely on an assumption of weak nonlinearities which allows to consider only lowest orders of a Taylor expansion about the equilibrium (2).

Amplitude equations have been worked out in the quasi adiabatic approximation, by means of an expansion of the solution of (1) over the eigenvector basis, truncated to a few modes (Takeuti and Aikawa 1981) and, under the assumption of slowly varying amplitudes, making use of averaging techniques (Dziembowski 1982).

One actually is correct in considering only a few modes but their number and their choice are not arbitrary. The normal form theory states (Guckenheimer and Holmes 1983) that, among the set of linear eigenmodes of the system, only the marginally stable and unstable ones (i.e. $|\kappa/\omega| << 1$) do determine the dynamics, which then takes place on a subspace of the phase space. The other (slave) modes only slightly distort this subspace and only indirectly contribute to the saturation of the marginally unstable modes. It is then possible to eliminate at each nonlinear order, one step at a time, all nonlinearities which do not play a determinant role.

Taking advantage of this theory, Coullet and Spiegel (1983) have achieved an elegant formalism which systematically derives the amplitude equations and their nonlinear coefficients. The main assumption, besides weak nonlinearities and marginality, is that the time dependence only occurs through the amplitudes of oscillation. Using a two-time asymptotic method with the assumption of weakly nonlinear, weakly nonadiabatic ($|\kappa|/\Omega < 1$) pulsations, Buchler and Goupil (1984) also developed an amplitude equation formalism in the specific framework of stellar nonadiabatic pulsations. Results from amplitude equations are found in accordance with those issued from hydrodynamical models and observations of classical Cepheids and RR Lyrae stars (Klapp et al. 1985; Buchler and Kovács 1987b; for a review, see Buchler 1989). In particular, they confirm the hypothesis of a 2:1 resonance for the Bump Cepheids. Chaotic amplitude equations result, for instance, from the nonlinear interaction between two marginal modes of which one is dynamically unstable (near vanishing frequency) (Argoul and Arnéodo 1984). Buchler and Goupil (1988) have suggested this mechanism, which generates intermittent oscillations, to explain irregular behavior of some supergiant stars.

It is out of the present scope, to detail these formalisms any further. We consider rather a system whose linearization gives rise to one single marginal radial mode, the other modes acting as slaves modes. In this case, saturation can already occur at cubic nonlinear order. The pulsation is then described, to first order, by: $\mathbf{z}(m,t) = a(t)\xi(m) + \mathrm{cc}$ where cc stands for complex conjugate; ξ and $\sigma = i\Omega + \kappa$ are the linear eigenvector and eigenfrequency of the marginal mode. The complex amplitude a obeys the amplitude equation:

$$\frac{da}{dt} = \sigma a + Qa|a|^2 \tag{13}$$

Setting $a = A\ \exp(i\phi)$ and $Q = R_eQ + iI_mQ$, (13) becomes:

$$\frac{dA}{dt} = (\kappa + R_eQA^2)\ A \equiv \kappa_{eff}A \tag{14a}$$

$$\frac{d\phi}{dt} = \Omega + I_mQA^2 \equiv \Omega_{NL} \tag{14b}$$

where κ_{eff} is the effective growth rate and Ω_{NL}, the nonlinear frequency of pulsation. One notes from (14) that, depending on the sign of R_eQ, cubic nonlinearities will enhance the exponential increase of the amplitude of an unstable mode if $\kappa_{eff} > 0$ or will inhibit it if $\kappa_{eff} < 0$. Only for $\kappa_{eff} = 0$ which requires $R_eQ < 0$, the amplitude reaches a finite, constant value, leading to a (limit cycle) periodic oscillation. The fixed point and the nonlinear frequency then are

$$A_{lc}^2 = \frac{\kappa}{|R_eQ|}; \qquad \Omega_{NL} = \Omega\left(1 + \frac{\kappa}{\Omega}\frac{I_mQ}{|R_eQ|}\right) \tag{15}$$

If in addition to, say, an unstable fundamental mode, an overtone is linearly unstable, it will be nonlinearly damped if its effective growth rate $\kappa_{eff\,o} = \kappa_o + R_eQ_oA_o^2 + R_eT_{of}A_f^2 < 0$, where o and f mean overtone and fundamental. This will happen if R_eQ_o and/or R_eT_{of} are sufficiently negative. Thus, the signs of the nonlinear coefficients Q, T determine the selection and saturation mechanisms. Either computed (Klapp et al. 1985) or empirically evaluated by fitting hydrodynamical results (Buchler and Kovács 1987b; Kovács and Buchler 1989), these coefficients have generally been found to be negative. This can be explained by an investigation of their physical contents in a way that we sketch below. We will make a number of assumptions, some of them necessary to be able to proceed and get a tractable expression, some others for shortness to keep up with the main topic of this paper. Using the formalism of Buchler and Goupil (1984), (see also Buchler 1985), the general expression of the self saturation coefficient Q in (13) involves two terms, $Q = Q^q + Q^c$. The contribution Q^q represents the interaction of the marginal mode and the slave modes. We expect it to be negative since stable modes tend to sip energy from the unstable marginal mode(s), thereby contributing to saturation. We henceforth assume $R_e Q^q < 0$ and restrict the discussion to the simpler coefficient Q^c, which expresses the nonlinear self-interaction of the marginal mode. From this last contribution, we consider only one of the two sources of nonlinearities, i.e. the nonlinear dependence of pressure and gravity on radius and, for shortness though they are not negligible, disregard nonlinearities of pressure and gravity due to the nonlinear variations of the radius itself. The expression of Q^c, in a continuous lagrangean description, is given by:

$$Q^c = \int_M dm\ \tilde{\xi} \cdot \mathcal{N}_3(\xi, \xi, \xi^*) = \int_M dm\ \left(\tilde{\xi}_r \mathcal{N}_{3r} + \tilde{\xi}_v \mathcal{N}_{3v} + \tilde{\xi}_s \mathcal{N}_{3s}\right) \tag{16}$$

where $*$ means complex conjugate. Subscripts r, v, s indicate the components of $\mathcal{N}_3$ and are symmetrized with respect to their arguments; ξ is the linear eigenvector of the marginal mode and $\tilde{\xi}$, the associated normalized adjoint eigenvector.

From (A2), the first term of the integrand in the second equality of (16) is always zero. By definition of a marginal mode, $|\kappa/\Omega| << 1$ i.e. the pulsation is globally weakly nonadiabatic. Typically, for a classical Cepheid or an RR Lyrae model, one has $|\kappa/\Omega| \sim 10^{-3} - 10^{-2}$. We will then neglect all terms of order $O(|\kappa/\Omega|)^2$ or higher. The assumption of marginality does not require the stellar interior to be locally quasiadiabatic. We, however here, assume local quasiadiabaticity i.e. $\mu << 1$ (see II-2), which has proven to be fruitful in the discussion of instability mechanisms (Cox 1980a). In this approximation, we use adiabatic eigenvectors but keep the complex eigenfrequency $\sigma = \kappa + i\Omega$. Then, ξ_r is real, normalized to 1 at the surface, ξ_s is given by (7c). From the hermiticity of the linear operator $\mathbf{L}$ in (6), one has $\tilde{\xi}_v = \bar{r}^2 \xi_r$ for the component $\tilde{\xi}_v$ of the adjoint eigenvector. The adjoint eigenvector must be normalized (i.e. $\tilde{\xi}_k \rightarrow \tilde{\xi}_k / N_k$) such that $\int_M dm\ \tilde{\xi}_k \cdot \xi_j = \delta_{kj}$. The norm (9) is then given by

$$N = 2\sigma \int_M dm\ \bar{r}^2 \xi_r^2 + \int_M dm\ \tilde{\xi}_s \xi_s = 2\sigma J + \int_M dm\ \tilde{\xi}_s \xi_s \tag{17}$$

In the adiabatic approximation, the last term in (17) and in the expression (16) do not exist. In the quasiadiabatic approximation, they are of order μ and must be taken into account. They are expected to play an important role in particular in the ionization zones since they express second and third order variations of the luminosity and temperature and of opacity. Nevertheless, we shorten the calculation here by neglecting both of them. The discussion of the remainder of this section should then be regarded as only indicative of the type of physical insight amplitude equations can provide. With these limitations in mind, we proceed with evaluating Q^c which reduces to:

$$Q^c = \frac{1}{2\sigma J}\int_M dm\ \bar{r}^2 \xi_r\ \mathcal{N}_{3v}(\xi,\xi,\xi^*) \tag{18}$$

The calculation, detailed in App. B, gives:

$$Q^c = -\kappa \Lambda_r + i\Omega\Lambda_i + O(\frac{\kappa}{\Omega})^2 \tag{19}$$

Inserting (19) into (15), assuming $R_eQ^q < 0, \kappa > 0$, one gets:

$$A^2_{lc} = \frac{1}{\Lambda_r + |R_eQ^q|/\kappa} \tag{20a}$$

$$\Omega_{NL} = \Omega(1+\Delta);\quad \Delta = \frac{\Lambda_i + I_mQ^q/\Omega}{\Lambda_r + |R_eQ^q|/\kappa} \tag{20b}$$

From (20a), a positive Λ_r contributes to the existence of a finite amplitude limit cycle. For $\Lambda_r < -|R_eQ^q|/\kappa < 0$, no saturation results from pressure variations at a cubic nonlinear order. The intermediate case, $-|R_eQ^q|/\kappa < \Lambda_r < 0$, causes an increase of the amplitude of the limit cycle. For the discussion, we neglect the derivatives of ξ_r with respect to m, which for a fundamental mode does not alter the qualitative results. From App. B, one then has:

$$\Lambda_r = \frac{3}{2}\int_M dm\ \xi_r^2\ \lambda_r(m);\qquad \Lambda_i = \frac{3}{2}\int_M dm\ \xi_r^2\ \lambda_i(m) \tag{21}$$

$$\lambda_r(m) = (9x + 1/3)\Phi_1(m) - 9y\Phi_2(m) \tag{22a}$$

$$\lambda_i(m) = (9x + 1/3)\ \Phi_1(m) \tag{22b}$$

Quantities entering (22) are defined in App. B. The first term in (22a) is due to adiabatic variations of pressure. The second term in (22a) represents the quasi adiabatic contribution to pressure variations. Recalling that κ is given by (11), $\Phi_2(m)$ expresses the local relative variation of the growth rate.

In the stellar interior, $\Gamma_1 \sim \Gamma_3 \sim 5/3$ are approximately constant i.e. $y \sim 1$ and x reduces to $(c_s/\Omega\bar{r})^2$. The sound speed is far larger than the velocity $\Omega\bar{r}$ i.e. $9x + 1/3 >> 0$. Since $\Phi_1(m) > 0$, the first term in (22a) is positive. Due to radiative damping, the interior is (linearly) stabilizing (Cox 1980a) i.e. $k(m)$ and $\Phi_2(m)$ (see B29, B30) are negative. Then, λ_r is positive in the interior and gravity and pressure variations in this region contribute to saturation.

The sound speed however decreases rapidly toward the surface while the velocity $\Omega\bar{r}$ increases. Therefore, $9x + 1/3$ becomes less possitive toward the surface (Fig. 3).

Moving up toward the surface, one encounters the helium partial second ionization zone. The decrease in x (B29) is enhanced there because of the sharp drop

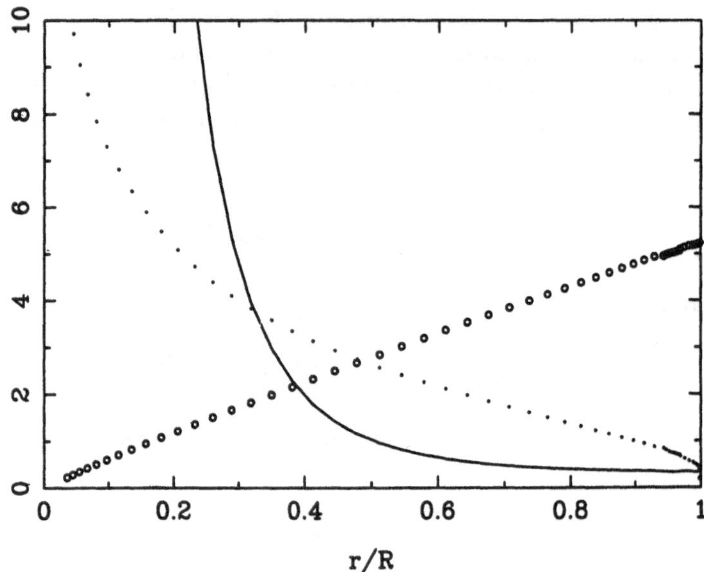

Fig. 3. The velocity $\Omega\bar{r}/R$ (dots) increases while $3c_s/R$ (open circles) decreases toward the surface and $9x + 1/3$ (full curve) becomes less positive (see text).

of the adiabatic exponents Γ_1 and Γ_3 below 5/3 (Fig. 1). When Γ_1 reaches the vicinity of 4/3, the sign of x strongly depends upon the derivatives of Γ_1 with pressure, i.e. on the equation of state in stellar models. On the other hand, $\Phi_2(m)$ becomes positive in the partial ionization zones where the κ and γ mechanisms operate. With $9x + 1/3$ less positive and $\Phi_2(m) > 0$ in this zone, λ_r adopts the sign of the dominant contribution and may become negative there.

The net result for the sign of Λ_r depends on the importance of the weight of the different zones relative to each other. Since the mode is unstable, the effect of the destabilizing (outer) regions wins over the stabilizing (inner) ones to produce $\kappa > 0$. This effect is enhanced by the weight $\xi_r^2 dm$ which tends to favor the outer layers and one must expect $\Lambda_r < 0$ resulting in an increase of the amplitude of the limit cycle.

The nonlinear frequency (22b) imposes the ratio of nonlinear to linear periods of the limit cycle: $P_{NL}/P_L = (1 + \Delta)^{-1}$. Nonlinear periods are larger than linear ones for $\Delta < 0$ but the ratio decreases with increasing Δ. In (22b), $(9x + 1/3)$ is positive i.e. $\lambda_i > 0$. Therefore, $\Lambda_i > 0$ gives a positive contribution to Δ (the denominator is positive for a limit cycle to exist) which tends to increase P_{NL}/P_L. The net sign of Δ, however, depends on the quantity $I_m Q^q$.

Nonlinear periods of most hydrodynamical models of classical Cepheids and RR Lyrae stars are usually found slightly longer than the linear ones, which implies that Δ must be negative and small in these models. Since the sign of $I_m Q^q$ is unknown at the moment, we assume it negative to allow a negative Δ as expected from stellar models. Stellingwerf (1975) remarked that in sequences of RR Lyrae models with L, M held fixed, the ratio P_{NL}/P_L decreases with decreasing effective temperature while the radius R increases and $\Omega \propto P_L^{-1}$ decreases. Since $\Omega \sim \pi < c_s > /R$, we admit that the stellar structure and as a consequence c_s are not much affected in these sequences. Then $x \sim (c_s/\Omega\bar{r})^2$, λ_i and thus Λ_i increase with Ω. On the other hand, Λ_r probably is less affected by the increase of x due to its additional quasi adiabatic contribution. Therefore, one expects that Δ increases with decreasing Ω and, though very crudely, we recover the tendency of Stellingwerf 's models i.e. the ratio P_{NL}/P_L decreases with increasing linear period. Clearly however, the sign of Q^q and its behavior with Ω have to be also investigated.

We recall that we have oversimplified the discussion for the sake of place and simplicity. Part of this oversimplification will be removed in a work to be published elsewhere. The final remark concerns the importance of the outer layers which are even more crucial in the nonlinear domain than in the linear one (the weight is in ξ_r^4 instead of ξ_r^2). Nonadiabatic contributions of the very outer layers can be qualitatively studied by means of an extreme nonadiabatic approximation. A quantitative analysis however must await a fully nonadiabatic numerical treatment.

II.5 Simplified Models

When the linear assumption is removed from Baker's model (II-2), the resulting nonlinear one zone model retains the main qualitative features of the finite amplitude pulsation of classical variables and probes the effects of some parameters upon phase lags and shapes of light curves (Stellingwerf 1972). Similarly, potential mechanisms suspected for driving stellar chaos can be tested with one zone models.

II.5.1 A Simple Nonlinear Model for Stellar Pulsations

When Eq. (12) is used to model a layer in a partial ionization zone, the physical behavior can be complemented by taking into account the variation of Γ_1 with the ionisation degree y. This was investigated by Buchler and Regev (1982) and Auvergne and Baglin (1985). Note that $\Gamma_1 < 4/3$ for $y = 0.5$ (Cox and Giuli 1968). In a crude way, but preserving the trend of Γ_1 (Fig. 1), it can be written as (Auvergne and Baglin 1985):

$$3\Gamma_1 - 4 = \gamma_1(1+\gamma_2)x^2; \quad \gamma_1 = 3\Gamma_{1(y=0.5)} - 4; \quad \gamma_2 = \frac{3}{2}\frac{\nu\eta^2/4}{(3+\nu\eta^2/4)^4} \tag{23}$$

where ν is the abundance by number of the ionizing element and $\eta = \chi/kT$ with χ, the ionization potential. The relative radius variations then obey:

$$\dddot{x} + \ddot{x} + \lambda(1+\alpha x^2)\dot{x} + \beta x = 0 \tag{24}$$

where the coefficients λ, α, β depend on γ_1, γ_2. This equation is a particular case of the equation studied by Moore and Spiegel (1966) and Baker et al. (1971) in a simplified model of radiating convective elements. System (24) exhibits chaotic oscillations for some ranges of parameter values. A Poincaré section, intersection of the trajectory with the plane $x = 0$ in phase space, clearly reveals the fractal structure of the attractor (Fig. 4a) after a suitable change of coordinates. In this model, chaos is due to a dynamical instability ($\Gamma_1 < 4/3$), which creates a saddle fixed point at the origin as in the Lorenz (1963) system.

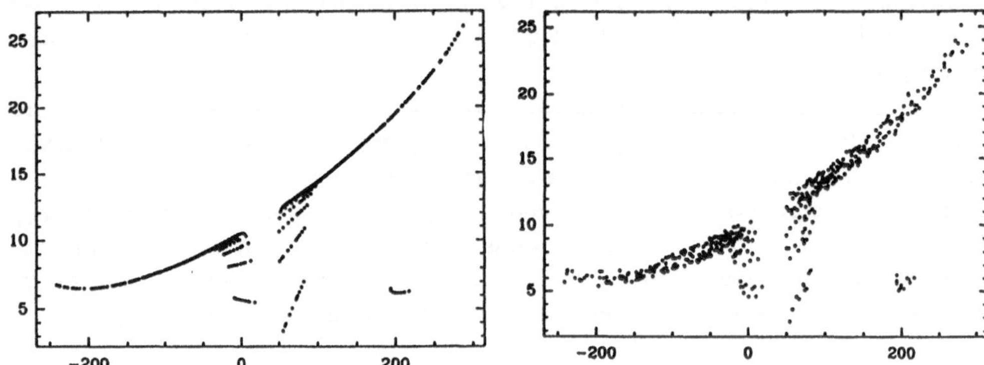

Fig. 4a. Poincaré section of a chaotic solution of (12).

Fig. 4b. Same with 1% noise added to the signal.

Saitou et al. (1989) consider a saturation arising from an ad hoc nonlinear dependence of the opacity exponents upon radius and pressure. In this case, a period doubling cascade, following a Hopf bifurcation (vibrational instability), generates a chaotic attractor whose topology ressembles the topology of the Rössler (1976) attractor.

II.5.2 A Nonlinear Iterative Map

An iterative map, mimicking stellar pulsations, has been worked out by Perdang (1989). The derivation of this map can be summarized in four steps.

a) A Taylor expansion of System (1) about the equilibrium (2);

b) The linear eigenvalue problem is assumed to be solved. Expansions of variables (i.e. $\sum_j q_j(t)\xi_j(m)$) over the basis of eigenvectors $\xi_j(m)$ are inserted into the nonlinear system. This yields an infinite number of ordinary differential equations for the amplitudes $q_j(t)$ after a space averaging has been performed.

c) One single mode is supposed unstable. The dynamics of the other (stable) modes are assumed to be enslaved to the main oscillation generated by the unstable (dominant) mode. This leads to an evolution equation for the unstable mode amplitude.

d) The last step consists in discretizing this evolution equation at times $n\Pi$ where Π is the period of the oscillations and n, an integer. The resulting quadratic map is

$$q(n+1) = \left((1+k)q(n) + q^2(n) + bq^{*2}(n) + c|q(n)|^2\right) \; K(q(n)) \qquad (25)$$

with q^* is the complex conjugate of q. The function $K(q)$ is a fourth order function in $|q|$, which prevents against a divergence of the solutions. The physics of the system is entirely included in the coefficients k, b, c.

With the linear coefficient k as the control parameter, numerical studies of (25) reveal a rich set of regular (nT-periodic, quasiperiodic) solutions and irregular (chaotic) ones. All three routes to chaos occur. Boundaries of the attracting sets are fractal, which leads the author to suggest that the boundary of the instability strip could be fractal and that nearby objects in the Herzsprung Russell diagram could present very different dynamics.

III. Chaotic Stars: How to Find Them, if Any?

In dealing with pulsating stars, searches for chaos must take into account some peculiarities of astrophysical data. The dynamics is generally known through one single observable, the velocity or more often luminosity variations. These time series are unterrupted with gaps due to day-night alternations and weather conditions, are time limited and noisy (S/N < 500).

A convenient way to determine whether a dynamical system is chaotic is to follow its transition to chaos when varying a control parameter such as, for instance, Reynolds or Rayleigh numbers in experimental studies of turbulence. The effective temperature or the mass have been chosen as control parameters in sequences of pulsating models (Kovacs and Buchler 1988; Aikawa 1987). Unfortunately, such a control upon the bifurcations of real stars is of course unavailable and transitions to chaos will have to be identified in other ways.

III.1 Phase Space Reconstruction

For a dissipative dynamical system known through one single time series, $s(t)$, the set of vectors $x(t) = \{s(t), s(t+2\tau), ..., s(t+(n-1)\tau\}$, where n is the dimension of the Euclidian space chosen for the reconstruction, is constructed with an arbitrary time delay τ. This process is an embedding i.e. the original and embedded sets are related through a diffeomorphism and the reconstructed attractor is topologically equivalent to the actual one (Takens 1980). This result is independent of the delay $\tau \neq 0$. However, if τ is too small, then $s(t) \sim s(t+\tau) \sim ... \sim s(t+(n-1))\tau)$ and all points remain close to the first bissector. Experience shows that the best choice for τ is of the order of 30% of the characteristic time scale of the data set.

A better visualisation of the attractor is provided by a singular value analysis (Broomhead and King 1986). Applied to pulsating white dwarfs, results suggest that physical quantities such as radius, temperature and velocity variations, can be qualitatively extracted from observed light curves, when the amplitude is not too large (Auvergne 1988).

These processes allow attractor reconstructions from time series with gaps. However, when considering relative variations, caution must be exercized in the evaluation of the mean value of each run in the data reduction.

Once, the attractor is reconstructed, Poincaré sections and return maps provide information on its structure, hence on the underlying dynamics. Unfortunately, noise in the data acts as a diffusive process in the embedding space. As an illustration, a one percent white noise, mimicking the atmospheric fluctuations, when added to the solution of (24), swaps out the fractal structure (Fig. 4b), the attractor broadens and fine structures are lost. Characteristics of the attractor are then affected to some extent by these perturbations.

III.2 Correlation Integral

Roughly speaking, the dimension of a set is related to the amount of information needed to locate a point on it. Several 'fractal dimensions' have been defined, depending on the nature of the desired informations (Farmer et al. 1983). Most commonly used, the correlation dimension is computed from the correlation integral as defined by:

$$C(l) = \lim_{N\to\infty} \frac{1}{N(N-1)} \sum_{i,j=1}^{N} \theta(l - \|x_i - x_j\|) \tag{26}$$

(Grassberger and Procaccia 1983), where θ is the Heavyside function. $C(l)$ counts the number of pairs of points such that $l < |x_i - x_j|$ (Fig. 5). For small values of l, one shows that $C(l) \sim l^D$, with D the attractor dimension. Since it depends on the value of D to be calculated, the dimension n of the embedding space is unknown. Therefore, $C(l)$ is computed for increasing values of n until the slope of $C(l)$ becomes independent of n, at which point one has reached the dimension of the set, for small l. $C(l)$ is independent of the choice of the norm.

Noise and shortness of the time series drastically affect the result of such a computation (Smith 1987). Due to the diffusive effect of the noise, at length scales l smaller than a value related to the variance of the noise, the slope of $C(l)$ generally sticks to the dimension of the embedding space (Fig. 5). In fact, the correlation dimension is related to the spectral properties of the noise. Osborne and Provenzale (1989) have shown that, for a coloured noise with a spectral power law $p(\omega) \propto \omega^{-\alpha}$, the correlation dimension decreases with increasing α. For a N dimensional noisy curve, with α between 1 and 2.5, the correlation dimension is found between $D_C > 10$ to $D_C = 1.4$ whereas the corresponding theoretical values lie between 10 and 1.6. Coloured noises of this kind with an exponent $\alpha \sim 1$, due to the atmospheric scintillation and transparency fluctuations, often damage atmospheric observations of pulsating stars. With such a power law, the noise does not influence the computed dimension of a low dimensional attractor whose fractal structure is such that the dimension is already reachable at intermediate length scales.

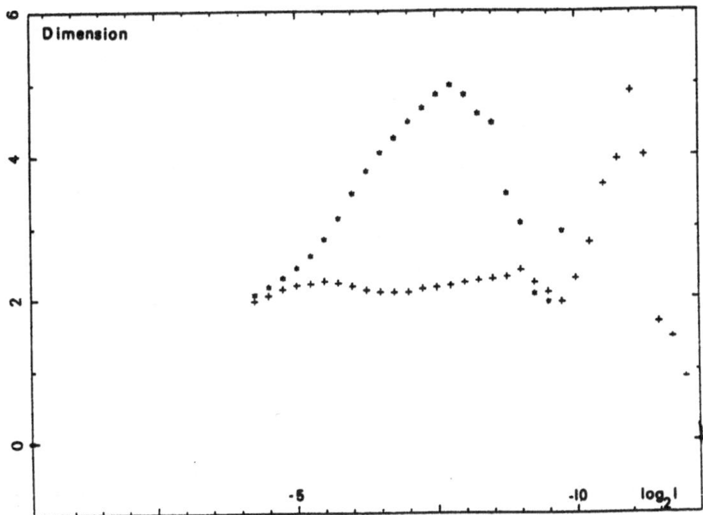

Fig. 5. Slope of $\log C(l)$ computed for a solution of Eq. (24). Without noise (crosses), with 1% noise (stars). The dimension of the attractor is 2.08.

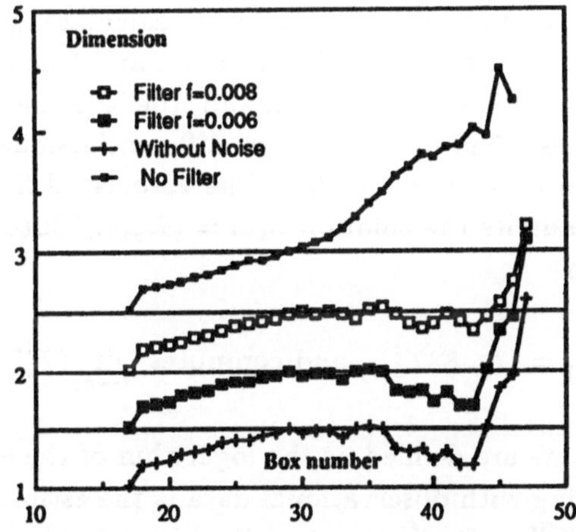

Fig. 6. Slope of the correlation integral for a torus in four cases: original data (crosses), with 2% noise (points), filtered with two differents cutoff frequencies of the lowpass filter (boxes).

III.3 Wavelet Analysis of Fractals

For a multifractal object, the scaling invariance properties are not identical everywhere on the attractor which is then characterized by generalized fractal dimensions and the spectrum of singularities of the corresponding invariant measures (Mandelbrot 1974; Hentschel and Procaccia 1983). Arnéodo et al. (1988) have shown that a multiscale technique, the wavelet transform (Grossmann et al. 1988), is an efficient tool to study the local self similarity of multifractal measures lying on Cantor sets, in spatially locating the different singularities and estimating their strengths. It is possible, for instance, to discriminate between uniform and non uniform Cantor sets. Transitions to chaos also are conveniently analyzed with such a technique.

The amount and quality of data necessary for a reliable investigation of this kind are expected to be of the same level as required by other characterization procedures, preventing an immediate application of the wavelet analysis as a space-scale technique to observational light curves. Nevertheless, lowering our demands, qualitative informations on the nonlinear behaviors of pulsating light curves can still be gained in considering the wavelet transform as a time-frequency analysis (Goupil et al. 1990).

III.4 Lyapunov Exponents

For a dissipative chaotic system, the volume in phase space always shrinks under the action of the flow but, while it contracts in some directions, it stretches in others. The Lyapunov exponents quantify this process. For regular (periodic, quasi periodic) variations, all exponents are negative or null whereas the exponential divergence of trajectories on strange attractors gives rise to at least one positive exponent.

Of the two methods that have been proposed to calculate Lyapounov exponents from experimental data (Wolf et al. 1985; Eckmann and Ruelle 1982; Eckmann et al. 1986), the method proposed by Eckmann and Ruelle is the most efficient. Let a dynamical system be described by an ordinary differential equation: $\dot{x} = F(x)$ and the solution be written as $x(t) = f^t(x(0))$. The associated linearized equation is $\dot{u} = (D_{x(t)}F)u$ which admits the solution $u(t) = (D_{x(0)}f^t)u(0)$. One then defines the tangent map T_x^t by:

$$T_x^t = \frac{\partial f_i^t}{\partial x_j} \text{ or } \frac{d}{dt}T_{x(0)}^t = (D_x F)T_{x(0)}^t \text{ and computes } \lim_{t\to\infty}(T_x^{t*}T_x^t)^{1/2t} = \Lambda_x. \quad (27)$$

The Lyapunov exponents are defined as the logarithm of the eigenvalues of Λ_x.

A difficulty in dealing with observational data is the estimation of the tangent map. The linear map $T_x^{\delta t} = D_x f^{\delta t}$ is determined by a least square fit. Successive matrices $T_x^{n\delta t}$ are computed. In the last step, one determines K upper triangular matrices R^m and the Lyapounov exponents are given by $\sum_{m=0}^{K-1} \log R_{kk}^m = \lambda_k$ up to a normalization factor.

For systems known through their differential equations, this method can provide all exponents. When the asymptotic behavior (attractor) is the only information, as is the case for stellar data, one has access to positive Lyapunov exponents only. However, negative exponents can also be obtained when transients are present, if intrinsic stochasticity perturbes the system, for instance. In this case, more information on the system is accessible (see Fig. 11).

The exponents λ_j being sorted in decreasing order, a Lyapunov dimension is defined as $D_\Lambda = k + c(k)/|\lambda_{k+1}|$, with $c(j) = \sum_1^j \lambda_j$, with the integer k such that $c(k) > 0$ and $c(k+1) < 0$. Kaplan and Yorke (1979) have conjectured that in *general*: $D_H = D_\Lambda$. The equality holds in many cases but counter examples are known.

It has been reported by Conte and Dubois (1987) that noise reduces the absolute value of Lyapounov exponents. This effect is particularly important for positive exponents because they generally are small. Because of its diffusion effect, noise mixes nearby trajectories. Only when trajectories have diverged enough on a strange attractor, can this divergence be detected but then the estimation of the tangent map becomes imprecise.

Badii et al. (1988) have stressed that a low pass filtering of the data introduces an extra positive Lyapounov exponent, if the frequency cutoff is lower than the smallest exponent. From the conjecture of Kaplan and Yorke, this should affect the computed dimension of a filtered chaotic attractor. Like a noise, chaos indeed produces a continuous spectrum. A filter therefore suppresses significant time scales of the chaotic signal and as a result, the computed dimension depends on the frequency cutoff. On the contrary, a quasiperiodic system has a line spectrum. If all the peak frequencies are lower than the frequency cutoff, we then expect that the dimension will remain independent of it. As a confirmation, Fig. 6 shows the dimension of a 2D-torus, computed with the correlation integral (26). A white noise has been added to the signal and a low pass filter applied to it, with two different cutoff frequencies. The dimension is found to be independent of the frequency cutoff. This property provides a convenient criterion to discriminate between quasiperiodic and chaotic dynamics (Auvergne 1991, in prep.).

Finally, J. Perdang (1990a) suggests to use pseudo periods, based on a modified definition of chaos, as more appropriate characteristic time scales of chaotic observational time series. These quantities should provide useful information, in much the same way than linear periods do. This has not yet been applied to real data.

III.5 Time Series Prediction

Predicting the future evolution of a deterministic system is always possible on a finite time interval (very short in chaotic regimes). A method based on a local estimation of the tangent map as for the Lyapunov exponents computation has been proposed (Farmer and Sidorowich 1988, and references therein). The future of each given point $x(t_{ref})$ on the attractor is generated by the flow $x(t_{ref}+1) = f(x(t_{ref}))$. The points $x(t_i)$, lying in a small ball centered on a given $x(t_{ref})$, and their known images $x(t_{i+1})$ by the flow are recorded to compute a locally averaged, linear approximation of the flow (tangent map) $x(t_{i+1}) = Lx(t_i)$ by a least square fit. Provided the balls are small enough, the future of x_{ref} is predicted as Lx_{ref}.

For a strange attractor, the prediction rapidly fails due to the exponential divergence, the error behaves as $\exp \lambda t$, where λ is the largest Lyapunov exponent. A study of the efficiency of the prediction on different parts of the attractor provides an estimate of the local strength of the divergence.

For reliable results, the attractor needs to be faithfully described (enough data) and only mildly noisy. Then, the method also yields an estimate of the minimal embedding dimension necessary to represent the concealed dynamics.

In the astrophysical context, this predictive method has been applied to sunspots (Kurths and Ruzmaikin 1990) and to white dwarf light curves (Serre et al. 1990).

IV. Do Chaotic Stars Really Exist?

Candidates for chaotic pulsations are the semi–regular variables which owe their names to the irregular behaviors of their light curves, many Mira stars whose periods from cycle to cycle irregularly vary, RV Tauri stars which exhibit irregular alternations of large and small light curve maxima. Of much shorter periods, most of non radial pulsating white dwarfs do also show complex light curves.

IV.1 RV Tauri and Mira Stars

Saitou et al. (1989) built first return maps, using the maxima of light curves, for RV Tauri stars. These maps show no smooth (1D) variations but rather complex patterns of clouds of points. In visually ordering the return maps of 5 stars, the authors found a trend of increasing complexity (scattering) with decreasing effective temperatures, reminiscent of the decrease in effective temperature which accompagnies the transition from order to desorder (to chaos) in their one zone model and in hydrocodes results from Kovács and Buchler (1988) for instance.

A singular value analysis of the light curve of the RV Tauri star, R Scuti, generates an attractor whose topology visually looks similar to the topology of the Rössler attractor (Kolláth 1990). However, a return map built in plotting the successive light curve minima shows a far larger scattering than return maps obtained for the Rössler system, which could indicate a higher dimension of the observed attractor (Fig. 7). The data were found too noisy to allow a determination of the correlation dimension of the attractor. Moreover, in the case of a high dimension (> 3), the number of cycles in the data set would be far too small for the correlation dimension to give a reliable result.

Kolláth (1990) estimates the embedding dimension to be about 4-5. An independent evaluation by means of a time prediction method gives a value of 5 or higher (Serre 1990, private communication), supporting the idea of a possible high dimensional chaos.

No regular patterns show up in return maps (Fig. 8) constructed with the time length between successive light curve cycles for Mira stars (Saijo et al. 1987; Saijo 1988; Perdang 1990a) or semi regular stars (Saitou et al. 1989; Zlodos 1990). Cannizzo et al. (1990) obtain noninteger but different values for the correlation dimension when evaluating the brightness and magnitude of light curves of three long period variables and therefore question the results as really indicative of chaotic variations. Indeed, noise probably dominates the length scale at which the dimension could be obtained as emphasized in Sect. III.2. Besides, the same limitation as above holds here, concerning the large number of required cycles when using high values of embedding space.

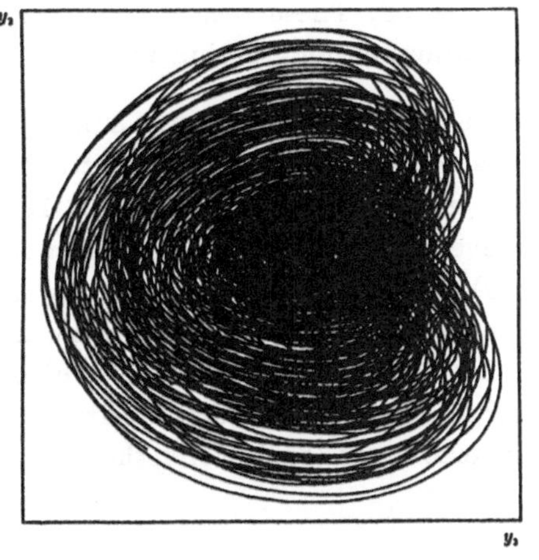

Fig. 7a. 2D projection of the reconstructed attractor of the light curve of a RV Tauri star, R Scuti (from Kooláth 1990).

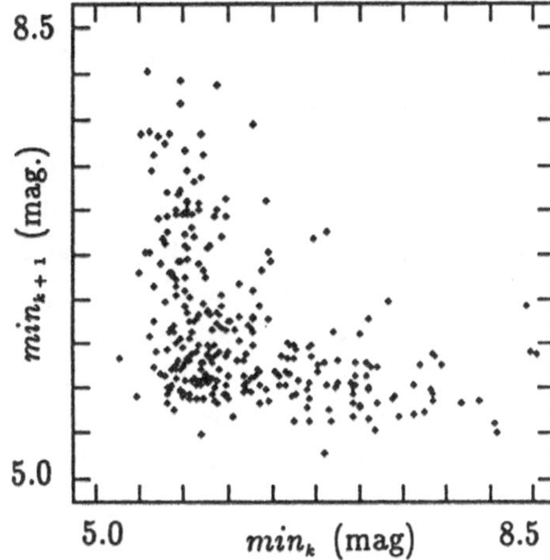

Fig. 7b. An associated first return map using light minima (from Kooláth 1990).

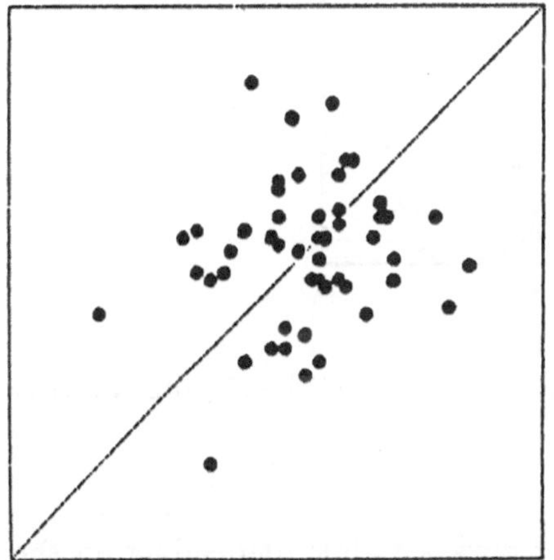

Fig. 8a. First return map, using light maxima, for the RV Tauri star, AC Her (from Saitou et al. 1989).

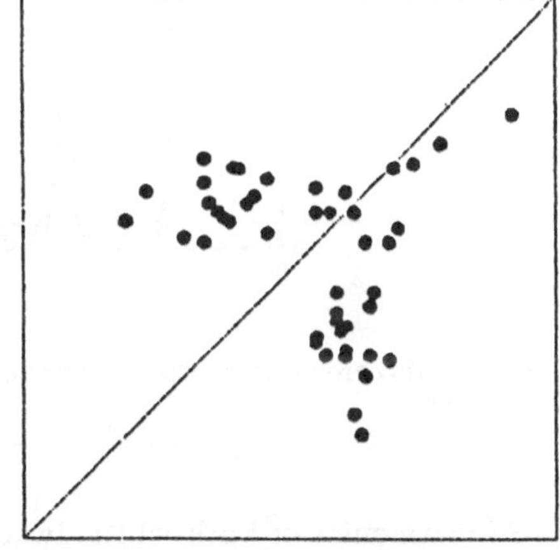

Fig. 8b. Same as Fig. 7a for the semi regular star, S Vul (from Saitou et al. 1989).

IV.2 Variable White Dwarfs

Despite the fact that they are faint objects, their complexities and their very short periods (ranging from two hundreds to a thousand seconds) make them attractive for seeking out stellar chaos. It is, indeed, possible to obtain about fourty periods in one night of observations. A correlation dimension of 2.5 has been obtained for the attractor of a DA white dwarf, GD 66 (Auvergne and Baglin 1986). A singular analysis (Auvergne 1988) and a time prediction method (Serre et al. 1990) suggest a low embedding dimension for the star PG1351+459. These results, however, suffer from the same uncertainties as the aforementionned long period variables.

Indication that the stars have possibly entered a chaotic regim through a period doubling cascade is the existence of subharmonics $\sim n\nu_0/2$ in the power spectrum of PG 1351 (Goupil et al. 1988) and $\sim n\nu_0/4$ in the power spectrum of G191 16. The equally spacing $\nu_0/4$ in the last case is best revealed with an autocorrelation of the power spectrum (Vauclair et al. 1989) (Fig. 9). Departures (as observed) from exact subharmonic ratios are actually expected for period doubling of autonomous systems undergoing period doubling cascades (Iooss and Joseph 1990) in which case they ought to measure, here, departures from adiabaticity. Such departures also occur when the system is in a region of inverse cascades, beyond the point where chaos sets in.

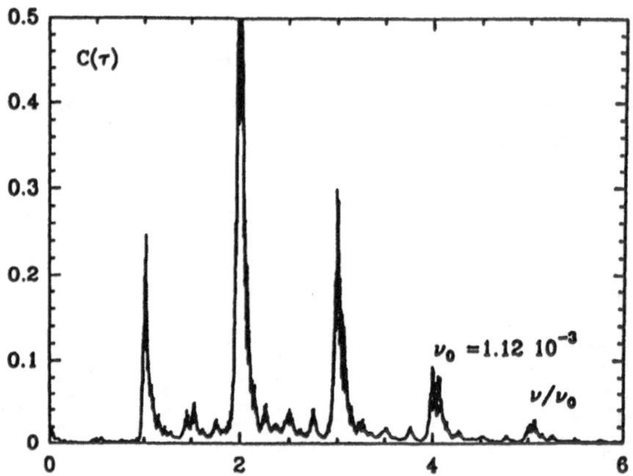

Fig. 9. Autocorrelation of the power spectrum of a ZZ Ceti star G191-16 (from Vauclair et al. 1989).

Subharmonics of both white dwarfs are not always visible, depending on the runs that are analyzed and, in the case of G191 16 which exhibits an amplitude modulation over a very long time scale ($\sim$ a month) compared to the main period (Auvergne et al. 1990b), subharmonics are detected when the amplitude is large, supporting the hypothesis of a nonlinear origin. These features indicate more complicated dynamics and would be observed, for instance if unstable modes, selected by a trapping mechanism (Winget et al. 1981), drive other modes through half integer resonances which generate subharmonics and, as found by Moskalik and Buchler (1990), chaos.

V. Conclusion

Simplified and hydrodynamical models, semi analytical approaches have grandly concurred to our current understanding of pulsating stars. Nonadiabatic effects and modal resonances are now known to play an important role in determining the dynamics and shaping the light and radial velocity curves. In the last decade, irregular stellar variations have intensively been examined within the framework of determinist chaos. Indeed, a chaotic attractor, in exploring a larger area of phase space, would provide more informations on the stellar structure than a periodic one.

From confrontations between observations and theoretical results, the following picture comes out: stars are dissipative dynamical systems whose sets of control parameters include at least the effective temperature T_{eff} and the luminosity to mass ratio L/M (Kovács and Buchler 1988). For most stars, values of these parameters prevent a vibrational destabilization and the star is in equilibrium (regarding the κ, γ mechanisms).

For other values of the control parameters, (as the star evolves, its structure and thus its control parameters vary), the structure is such that partial ionization zones enter the appropriate regions of the envelope and trigger a vibrational instability i.e. the star undergoes a Hopf bifurcation, growth rates of some modes become marginally positive ($\kappa \geq 0$). The oscillation amplitude of generally one mode increases to some extent until saturation by (low order) nonlinearities gives rise to a limit cycle oscillation.

For decreasing T_{eff} and $(L/M)^{-1}$, which places the stars close to the region of observed irregularity, increase of nonadiabaticity i.e. of the growth rates ($\kappa >> 0$) and of the limit cycle amplitude entail new nonlinear effects. Several internal resonances appear which drag additional modes into the dynamics. Partial ionization zones experiment larger motions, generating new or enhancing driving and dissipative processes. These mechanisms give birth to low dimensional chaos which, depending on the strength of L/M, arises through different universal scenarii and numerical simulations of light and velocity curves exhibit irregular, in some cases erratic oscillations.

The above picture presents the advantage of explaining, in a simple way, existence and location of regions of regular and irregular pulsations, observed in the upper part of the Hertzsprung-Russell diagram.

Use of characterization procedures, which derive from the theory of dissipative dynamical systems, gives some observational clues in favour of the existence of chaotic stellar pulsators though no definite conclusion can be drawn. The arguments against chaotic variations essentially rely on the complexity of the observed return maps since numerical simulations, so far, have provided almost one dimensional, nicely shaped first return maps. Most often, therefore, the complexity of observed return maps has been attributed to regular oscillations perturbed by noise.

Observational noise, at least in data of Mira and RV Tauri stars, however, is not strong enough to account for the large scatter of their first return maps. Actually,

Mattei et al. (1990) found that, if one wants to model Mira light curves with a periodic signal, a random part with an amplitude as large as half the amplitude of the deterministic signal must be added.

On the other hand, deterministic chaos can give rise to quite complex first return maps. In particular, a type III intermittency (Bergé et al. 1986), which proceeds from a subcritical bifurcation of a subharmonic limit cycle, shows unimodal nth return maps when n is the order of the subharmonic and first return maps can be quite complex. In stellar modelling, this transition to chaos has been observed in numerical experiments of the quadratic map (25) (Perdang 1990b). Recently, Aikawa (1990) studied a chaotic model of pop II Cepheid whose transition to chaos looks similar to a type III intermittent scenario but, according to the quoted author, follows a period doubling cascade unterrupted after two period doublings by a type I intermittency. The point, here, is that, while the fourth return map displays an almost one dimensional, simple shape, the structure of the first return map is much more entwined (Fig. 10). Similar processes are likely to occur in stellar pulsations and could explain some observed complex return maps. This may be more difficult to emphasize, however, since more points are required and noise may be more perturbing with increasing orders of the return map.

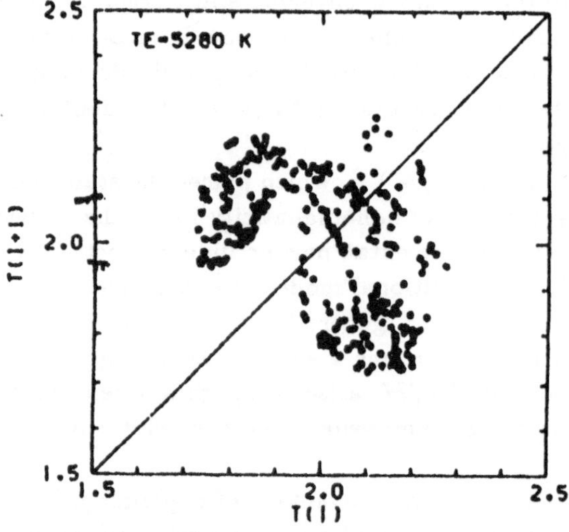

Fig. 10. First return map using kinetic energy maxima at expansion phases for a chaotic pop.II Cepheid model (from Aikawa 1990).

Alternatively, a complex first return map can arise from a high dimensional chaos. Its observational evidence would be more difficult to obtain but would explain, for instance, an embedding dimension as large or larger than 5 as estimated for some stars. Physical mechanisms involving a large number of degrees of freedom, as convection present in stellar envelopes for instance, could be instrumental in generating such a high dimensional chaos. This is emphasized by Perdang (1990a), who then suggests to account for such a large number of degrees by introducing an intrinsic noise in the stellar pulsation models, leading to a description in terms of stochastic differential equations. As an illustration, Perdang's map (25) is plotted in a chaotic case chosen from Perdang (1990) on Fig. 11a. Fig. 11b displays the same return map but with 10% white noise added at each iteration. Due to intrinsic noise, the structure of the map is more fully described as emphasized in Sect. III.4.

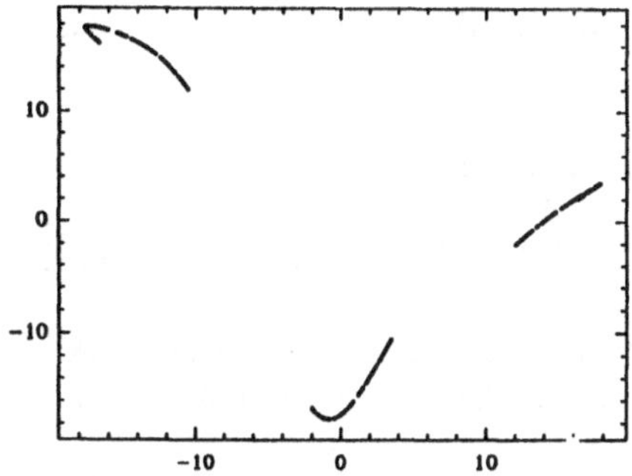

Fig. 11a. $x(i)$ versus $x(i+1)$ for the map derived in Perdang (1989) with parameter value $\epsilon = 0.1$, $\alpha = 2$, $k = 0.1$, $c_R = -0.001$, $b_R = b_I = c_I = 1$.

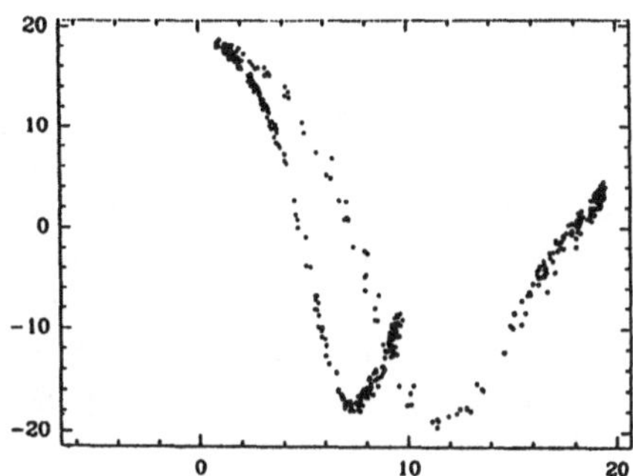

Fig. 11b. Same as Fig. 11a, with 10% intrinsic noise.

This encourages further investigations with appropriate simplified models, improved hydrocodes and semi analytical approaches. It is possible to complete our physical understanding of stellar transitions to chaos in studying amplitude equations which can be derived for nT-periodic solutions and quasiperiodic solutions, bifurcating from a T-periodic (Iooss and Joseph 1990). Eigenvalues (Floquet exponents) and eigenvectors of the Floquet matrix of system are used to build an approximate nT-periodic solution and associated amplitude equations. The period doubled (2T) limit cycle, for instance, results from the 'resonance' 2:1 between the limit cycle frequency and a Floquet exponent and admits an amplitude equation of cubic nonlinear order in amplitude (Iooss and Joseph 1990). Again fixed points and stability can be infered from these equations. This extension of the amplitude equation formalism has not yet been adapted to the astrophysical framework (in prep.)

A final word, concerning pulsating white dwarfs: many modes can be excited in a nonradial pulsator and their superposition can give rise to light curves which look irregular. Some white dwarfs have indeed been observed pulsating with several hundreds of frequencies. In most of these cases, a linear interpretation seems satisfactory (Winget 1990). For a few other white dwarfs, observational indications exist in favour of chaotic variations. The different possible origins of chaos described above in the radial case extend to the non radial one. Progress regarding this possibility must await – currently nonexisting – nonradial hydrodynamical models or further observations.

References

Aikawa, T. (1987): *Astrophys. Sp. Sc.* **139**, 281.
Aikawa, T. (1988): *Astrophys. Sp. Sc.* **149**, 149.
Aikawa, T. (1990): *Astrophys. Sp. Sc.* **164**, 295.
Antonello, E., Mantegazza, L. (1984): *Astron. Astrophys.* 133, 52.
Argoul, F., Arnéodo, A. (1984): *J. Mech. Th. Appl.* **241**, 118.
Arnéodo, A., Argoul, F., Elezgaray, J, Grasseau, G. (1988): in *Proc. Conf. Nonlinear Dynamics*, Ed. G.G. Turchetti.
Auvergne, M. (1988): *Astron. Astrophys.* **204**, 341.
Auvergne, M., Baglin, A. (1985): *Astron. Astrophys.* **142**, 388.
Auvergne, M., Baglin, A. (1986): *Astron. Astrophys.* **168**, 118.
Auvergne, M., Goupil, M.J., Baglin, A., Serre, T. (1990a): in *Fractals in Astronomy*, Ed. A. Heck, *Vistas Astron.* **33**, 399.
Auvergne, M., Chevreton, M., Belmonte, J., Vauclair, G., Dolez, N., Goupil, M.J. (1990b): in *Proc. of 7th Eur. Workshop White Dwarfs*, Ed. G. Vauclair (Kluwer Acad. Publ., Dordrecht), in press.
Badii, R., Broggi, G., Derighetti, B., Ravani, M., Ciliberto, S., Politi, A., Rubio, M.A. (1988): *Phys. Rev. Letters* **60**, 979.
Baker, N.H. (1966): in *Stellar Evolution*, Eds. R.F. Stein, A.G.W. Cameron, (Plenum Press, New York), p. 333.
Baker, N.H., Moore, D.W., Spiegel, E.A. (1971): *Quart. J. Mech. Appl. Math.* **XXIV**, 391.
Bergé, P., Pommeau, Y., Vidal, C. (1986): *Order within Chaos* (Wiley, New York).
Broomhead, D.S., King, G.P. (1986): *Physica D* **20**, 217.
Buchler, J.R. (1985): in *Chaos in Astrophysics*, NATO ARW **C 161** (D. Reidel Publ. Co., Dordrecht), p. 137.
Buchler, J.R. (1989): in *The Numerical Modeling of Nonlinear Stellar Pulsations, Problems and Prospects*, Ed. J.R. Buchler (Kluwer Acad. Publ., Dordrecht), p. 1.
Buchler, J.R., Regev, O. (1982): *Astrophys. J.* **263**, 312.
Buchler, J.R., Goupil, M.J. (1984): *Astrophys. J.* **279**, 394.
Buchler, J.R., Goupil, M.J. (1988): *Astron. Astrophys.* **190**, 137.
Buchler, J.R., Goupil, M.J., Kovács, G. (1987): *Phys. Letters* **126**, 177.
Buchler, J.R., Kovács, G. (1987a): *Astrophys. J.* **320**, L57.
Buchler, J.R., Kovács, G. (1987b): *Astrophys. J.* **318**, 232.
Buchler, J.R., Moskalik, P., Kovács, G. (1990): *Astrophys. J.* **351**, 617.
Cannizzo, J.K., Goodings, D.A., Mattei, J.A. (1990): *Astrophys. J.* **357**, 235.
Christensen-Dalsgaard, J. (1988): *Advances in Helio and Asteroseismology*, Eds. J. Christensen-Dalsgaard, S. Franden (D. Reidel Publ. Co., Dordrecht), p. 3.
Christy, R.F. (1966): *Astrophys. J.* **144**, 108.
Christy, R.F. (1967): *Astrophys. J.* **145**, 337.
Conte, R., Dubois, M. (1987): in *Proc. IVth. Workshop Nonlinear Evolution Equations and Dynamical Systems*, Ed. J. Leon (World Scientific, Singapore), p. 767.
Coullet, P., Spiegel, E. (1983): *J. Appl. Math.* **43**, 776.
Cox, J.P. (1980a): *Theory of Stellar Pulsations* (Princeton Univ. Press., Princeton).
Cox, J.P. (1980b): in *Cepheids: Theory and Observations*, Ed. B.F. Madore (Cambridge Univ. Press, Cambridge), p. 126.
Cox, J.P., Giuli, R.N. (1968): *Principle of Stellar Structures* (Gordon and Breach, New York).

Cvitanović, P. (1984): *Universality in Chaos* (Hilger, Bristol).
Dziembowski, W., 1982, Acta Astronomica, **32**, 147.
Eckmann, J.P., Ruelle, D. (1982): *Rev. Mod. Phys.* **57**, 617.
Eckmann, J.P., Oliffson Kamphorst, S., Ruelle, D., Ciliberto S. (1986): *Phys. Rev. A* **34**, 4971.
Fadeyev, Y.A., Fokin, A. (1985): *Astrophys. Sp. Sc.* **111**, 355.
Farmer, J.D., Ott, E., Yorke, J.A. (1983): *Physica D* **7**, 153.
Farmer, J.D., Sidorowich, J.J. (1988): in *Cter. Nonlinear Studies, Los Alamos Lab. Rep.* **LA-UR-88-901**, 277.
Fokin, A. (1990): *Astrophys. Sp. Sc.* **164**, 95.
Glasner, S.A., Buchler, J.R. (1989): in *The Numerical Modeling of Nonlinear Stellar Pulsations, Problems and Prospects*, Ed. J.R. Buchler (Kluwer Acad. Publ., Dordrecht), p. 109.
Goupil, M.J., Auvergne, M., Baglin, A. (1988) *Astron. Astrophys.* **196**, L13.
Goupil, M.J., Auvergne, M., Baglin, A. (1990): in *Proc. 7th Eur. Workshop White Dwarfs*, Ed. G. Vauclair (Kluwer Acad. Publ., Dordrecht), in press.
Grassberger, P., Procaccia, I. (1983): *Phys. Rev. A* **28**, 2591.
Grossmann, A., Kronland-Martinet, R., Morlet, J. (1988): in *Wavelets*, Eds. J.M. Combes, A. Grossmann, Ph. Tchamitchian (Springer–Verlag, Berlin), p. 2.
Guckenheimer, J., Holmes, P. (1983): *Nonlinear Oscillations, Dynamical Systems and Bifurcation Theory* (Springer–Verlag, New York).
Hentschel, H.G.E., Procaccia, I. (1983): *Physica D* **8**, 435.
Iooss, G., Joseph, D.D. (1990): *Elementary Stability and Bifurcation Theory* (Springer–Verlag, New York).
Kaplan, J.L., Yorke, J.L. (1979): *Commun. Math. Phys.* **67**, 93.
Karp, A.H. (1975): *Astrophys. J.* **199**, 448.
King, S.K., Cox, J.P., Eilers, D.D., Davey, W.R. (1973): *Astrophys. J.* **182**, 859.
Klapp, J., Goupil, M.J., Buchler, J.R. (1985): *Astrophys. J.* **296**, 514.
Kolláth, M. (1990): *Monthly Not. Roy. Astron. Soc.* **247**, 377.
Kovács, G., Buchler, J.R. (1988): *Astrophys. J.* **334**, 971.
Kovács, G., Buchler, J.R. (1989): *Astrophys. J.* **346**, 898.
Kurths, E.N., Ruzmaikin, A.A. (1990): *Solar Physics* **126**, 407.
Ledoux, P., Walraven, Th. (1958): *Handbuch der Physik* **51**, 353.
Lorenz, E.N. (1963): *J. of Atm. Sci.* **20**, 130.
Mandelbrot, B.B. (1974): *J. Fluid Mech.* **62**, 331.
Mattei, J.A., Cannizzo, J.K., Goodings, D.A. (1990): in *Confrontation between Stellar Pulsation and Evolution*, Ed. C. Cacciari, G. Clementi, *Astron. Soc. Pac. Conf. Series* **11**, 461.
Moore, D.W., Spiegel, E.A. (1966): *Astron. Astrophys.* **116**, 341.
Moskalik, P., Buchler, J.R. (1990): *Astrophys. J.* **355**, 590.
Osborne, A.R., Provenzale, A. (1989): *Physica D* **35**, 357.
Ostlie, D.A., Cox, A.N., Cahn, J.H. (1982): in *Pulsation in Classical and Cataclismic Variables*, Eds. J.P. Cox, C.J. Hansen (Joint Inst. Lab. Astrophys., Boulder), p. 297.
Perdang, J. (1985): in *Chaos in Astrophysics*, NATO ARW **C 161**, Eds. J.R. Buchler, J.M. Perdang, E.A. Spiegel (D. Reidel Publ. Co., Dordrecht), p. 11.
Perdang, J. (1989): in *The Numerical Modeling of Nonlinear Stellar Pulsations: Problems and Prospects*, Ed. J.R. Buchler (Kluwer Acad. Publ., Dordrecht), p. 333.
Perdang, J. (1990a): in *Rapid Variability of OB Stars*, ESO Workshop, Ed. D. Baade (ESO, Garching–bei–München), in press.

Perdang, J. (1990b): in *Ecole de Structure Interne* **2**, eds. A.M. Hubert, E. Schatzman (DASGAL, Obs. Paris), in press.
Petersen, J.O. (1973) *Astron. Astrophys.* **27**, 89.
Provost, J., Berthomieu, G. (1990): in *Ecole de Structure Interne* **2**, Eds. A.M. Hubert, E. Schatzman (DASGAL, Obs. Paris), in press.
Rössler, O.E. (1976): *Phys. Letters* **57A**, 397.
Saijo, K. (1988): *Bull. Nat. Sci. Mus., Tokyo, Ser. E* **14**, 1.
Saijo, K., Watanabe, M. (1987): *Bull. Nat. Sci. Mus., Tokyo, Ser. E* **10**, 1.
Saitou, M., Takeuti, M., Tanaka, Y. (1989): *Publ. Astron. Soc. Japan* **41**, 297.
Serre, T., Buchler, J.R., Goupil, M.J. (1990): in *Proc. 7th Eur. Workshop White Dwarfs*, Ed. G. Vauclair (Kluwer Acad. Publ., Dordrecht), in press.
Simon, N. (1988): in *Pulsation and Mass Loss in Stars*, Eds. R. Stalio, L.A. Willson (Kluwer Acad. Publ., Dordrecht), p. 27.
Simon, N. (1990): *Monthly Not. Roy. Astron. Soc.* **246**, 70.
Simon, N., Aikawa, T. (1986): *Astrophys. J.* **304**, 249.
Simon, N., Davis, C.G. (1983): *Astrophys. J.* **266**, 787.
Smith, L.A. (1987): *Phys. Letters* **133A**, 283.
Stellingwerf, R.F. (1972): *Astron. Astrophys.* **21**, 91.
Stellingwerf, R.F. (1975): *Astrophys. J.* **195**, 441.
Stobie, R.S. (1969a): *Monthly Not. Roy. Astron. Soc.* **144**, 461.
Stobie, R.S. (1969b): *Monthly Not. Roy. Astron. Soc.* **144**, 485.
Stobie, R.S. (1969c): *Monthly Not. Roy. Astron. Soc.* **144**, 511.
Stothers, N.R. (1981): *Astrophys. J.* **247**, 941.
Takens, F. (1980): in *Dynamical System and Turbulence*, Eds. D.A. Rand, L.S. Young (Springer–Verlag, Berlin), p. 366.
Takeuti, M. (1987): *Astrophys. Sp. Sc.* **136**, 129.
Takeuti, M., Aikawa, T. (1981): *Sci. Rept. Tokohu Univ. 8th Seriee* **2**, 106.
Tresser, C., Coullet, P., Arnéodo, A. (1980): *J. Phys. Letters* **41**, L243.
Tuchman, Y., Sack, N., Barkat, Z.K. (1978): *Astrophys. J.* **219**, 183.
Tuchman, Y., Sack, N., Barkat, Z.K. (1979): *Astrophys. J.* **234**, 217.
Unno, W., Osaki, Y., Ando, H., Saio, H., Shibahashi, H. (1989): *Nonradial Oscillations of Stars* (Univ. Tokyo Press, Tokyo).
Vauclair, G., Goupil, M.J., Baglin, A., Auvergne, M., Chevreton, M. (1989): *Astron. Astrophys.* **215**, L17.
Winget, D. (1990): in *Proc. 7th Eur. Workshop on White Dwarfs*, Ed. G. Vauclair (Kluwer Acad. Publ., Dordrecht), in press.
Winget, D., Van Horn, H.M., Hansen, C.J. (1981): *Astrophys. J.* **245**, L55.
Wolf, A., Swift, J.B., Swinney, H.L., Vastano, J.A. (1985): *Physica D* **16**, 285.
Zlodos, E. (1990): *Astrophys. Sp. Sc.* **165**, 111.

Appendix A

Only components of the vector operators in (5) useful in this paper are defined:

$$L_r = v; \quad L_v = -4\pi\bar{r}\frac{\partial\delta_1 P}{\partial m} + 4\frac{g}{r}\frac{\delta r}{r}; \quad L_s = -\frac{1}{c_v\bar{T}}\frac{\partial\delta_1 L}{\partial m} \tag{A1}$$

$$N_{2r} = N_{3r} = 0 \tag{A2}$$

$$N_{2v} = -4\pi\bar{r}\left(\frac{\partial\delta_2 P}{\partial m} + 2\frac{\delta r}{\bar{r}}\frac{\partial\delta_1 P}{\partial m}\right) - 2\frac{g}{\bar{r}}(\frac{\delta r}{\bar{r}})^2 \tag{A3}$$

$$N_{3v} = -4\pi\bar{r}\left(\frac{\partial\delta_3 P}{\partial m} + 2\frac{\delta r}{\bar{r}}\frac{\partial\delta_2 P}{\partial m} + (\frac{\delta r}{\bar{r}})^2\frac{\partial\delta_1 P}{\partial m}\right) + 4\frac{g}{\bar{r}}(\frac{\delta r}{\bar{r}})^3 \tag{A4}$$

Pressure variations of second and third order are obtained from the Taylor expansion of the equation of state $P = P(\rho, s)$. We neglect the radiation pressure. One has:

$$\frac{\delta_1 P}{\bar{P}} = \Gamma_1\frac{\delta_1\rho}{\rho} + \chi_t\frac{\delta s}{c_v} \tag{A6}$$

$$\frac{\delta_2 P}{\bar{P}} = \Gamma_1\frac{\delta_2\rho}{\rho} + \frac{1}{2}\alpha_1(\frac{\delta_1\rho}{\rho})^2 + [\frac{1}{2}\alpha_3(\frac{\delta s}{c_v})^2 + \alpha_2\frac{\delta_1\rho}{\rho}\frac{\delta s}{c_v}] \tag{A7}$$

$$\frac{\delta_3 P}{\bar{P}} = \Gamma_1\frac{\delta_3\rho}{\rho} + \alpha_1\frac{\delta_1\rho}{\rho}\frac{\delta_2\rho}{\rho} + \frac{1}{6}\alpha_4(\frac{\delta_1\rho}{\rho})^3$$
$$+ [\alpha_2\frac{\delta s}{c_v}\frac{\delta_2\rho}{\rho} + \frac{1}{2}\alpha_5\frac{\delta s}{c_v}(\frac{\delta_1\rho}{\rho})^2 + \frac{1}{2}\alpha_6\frac{\delta_1\rho}{\rho}(\frac{\delta s}{c_v})^2 + \frac{1}{6}\alpha_7(\frac{\delta s}{c_v})^3] \tag{A8}$$

where derivatives of the pressure are written in terms of usual thermodynamical quantities (Unno et al. 1989):

$$\Gamma_1 = \left(\frac{\partial LogP}{\partial Log\rho}\right)_s ; \ \Gamma_3 = \left(\frac{\partial LogT}{\partial Log\rho}\right)_s ; \ \chi_t = \left(\frac{\partial LogP}{\partial LogT}\right)_\rho \tag{A9}$$

and we have defined:

$$\alpha_1 = \frac{\rho^2}{P}\left(\frac{\partial^2 P}{\partial\rho^2}\right)_s = \Gamma_1\left(\Gamma_1 - 1 + \left(\frac{\partial\Gamma_1}{\partial LogP}\right)_s\right) \tag{A10}$$

$$\alpha_2 = \frac{c_v\rho}{P}\left(\frac{\partial^2 P}{\partial\rho\partial s}\right); \quad \alpha_3 = \frac{c_v^2}{P}\left(\frac{\partial^2 P}{\partial s^2}\right)_\rho \tag{A11}$$

$$\alpha_4 = \frac{\rho^3}{P}\left(\frac{\partial^3 P}{\partial\rho^3}\right)_s = \Gamma_1\ [(\Gamma_1 - 1)(\Gamma_1 - 2) + 3(\Gamma_1 - 1)\left(\frac{\partial\Gamma_1}{\partial LogP}\right)_s$$
$$+ \left(\frac{\partial\Gamma_1}{\partial LogP}\right)_s^2 + \Gamma_1\left(\frac{\partial^2\Gamma_1}{\partial LogP^2}\right)_s] \tag{A12}$$

$$\alpha_5 = \frac{c_v\rho^2}{P}\left(\frac{\partial^3 P}{\partial\rho^2\partial s}\right) = \chi_t\ [\Gamma_3(\Gamma_3 - 1) + (3\Gamma_3 - 2)\left(\frac{\partial\Gamma_3}{\partial LogT}\right)_s$$
$$+ \left(\frac{\partial\Gamma_3}{\partial LogT}\right)_s^2 + (\Gamma_3 - 1)\left(\frac{\partial^2\Gamma_3}{\partial LogT^2}\right)_s] \tag{A13}$$

$$\alpha_6 = \frac{c_v^2\rho}{P}\left(\frac{\partial^3 P}{\partial\rho\partial s^2}\right); \quad \alpha_7 = \frac{c_v^3}{P}\left(\frac{\partial^3 P}{\partial s^3}\right)_\rho \tag{A14}$$

Density variations are obtained from the Taylor expansion of the continuity equation (1c) which gives:

$$\frac{\delta_1\rho}{\rho} = -(3\frac{\delta r}{\bar{r}} + 4\pi\bar{r}^3\bar{\rho}\ \frac{d\delta r/\bar{r}}{dm}) \tag{A15}$$

$$\frac{\delta_2\rho}{\rho} = (\frac{\delta_1\rho}{\rho})^2 - 3(\frac{\delta r}{\bar{r}})^2 - 4\pi\bar{r}^3\bar{\rho}\ 2\frac{\delta r}{\bar{r}}\frac{d\delta r/\bar{r}}{dm} \tag{A16}$$

$$\frac{\delta_3\rho}{\rho} = 2\frac{\delta_1\rho}{\rho}\frac{\delta_2\rho}{\rho} - (\frac{\delta_1\rho}{\rho})^3 - (\frac{\delta r}{\bar{r}})^3 - 4\pi\bar{r}^3\bar{\rho}\ (\frac{\delta r}{\bar{r}})^2\frac{d\delta r/\bar{r}}{dm} \tag{A17}$$

Appendix B

The nonlinear self interaction of a marginal mode leads to the saturation coefficient (18) or:

$$Q^c = \frac{1}{2\sigma J}\int_M dm\ \bar{r}^2\xi_r\ [N_{3v}(\xi,\xi,\xi^*) + N_{3v}(\xi,\xi^*,\xi) + N_{3v}(\xi^*,\xi,\xi)] \tag{B1}$$

which involves expressions of the type

$$\int_M dm\ \bar{r}^2\xi_r N_{3v}(\xi,\xi,\xi) \tag{B2}$$

The cubic nonlinearity N_{3v} in (B2) is given by (A4) and taking into account (7a,b), is rewritten as:

$$N_{3v}(\xi,\xi,\xi) = -4\pi\bar{r}\ (2\xi_r\frac{d\delta_2 P}{dm} + \frac{d\delta_3 P}{dm}) + \sigma^2\xi_r^3 \tag{B3}$$

Inserting (B3) into (B2), integrating by part, noting that

$$\bar{\rho}\frac{d4\pi\bar{r}^3\xi_r}{dm} = -\frac{\delta_1\rho}{\rho}; \qquad \bar{\rho}\frac{d4\pi\bar{r}^3\xi_r^2}{dm} = -\left(\frac{\delta_2\rho}{\rho} - (\frac{\delta_1\rho}{\rho})^2\right) \tag{B4}$$

one obtains:

$$\int_M dm\ \bar{r}^2\xi_r N_{3v}(\xi,\xi,\xi) =$$

$$\int_M dm\ [\frac{-\bar{P}}{\bar{\rho}}\left(2(\frac{\delta_2\rho}{\bar{\rho}} - (\frac{\delta_1\rho}{\bar{\rho}})^2)\frac{\delta_2 P}{\bar{P}} + \frac{\delta_1\rho}{\bar{\rho}}\frac{\delta_3 P}{\bar{P}}\right) + \sigma^2\bar{r}^2\xi_r^4] \tag{B5}$$

provided that the boundary conditions satisfies:

$$\left[4\pi\bar{r}^3\bar{P}\xi_r(2\xi_r\frac{\delta_2\bar{P}}{\bar{P}} + \frac{\delta_3 P}{\bar{P}})\right]_0^M = 0 \tag{B6}$$

This condition is satisfied with the usual boundary conditions: $\bar{r}^3\bar{P} = 0$ for $m = 0$ and $\bar{P} = 0$ in $m = M$.

With ξ_r and $\delta_1\rho/\rho$ real in the quasi adiabatic approximation, the circular symmetrization in (B1), with (B5), yields:

$$Q^c = \frac{1}{2\sigma J}\int_M dm \; - \frac{\bar{P}}{\bar{\rho}}$$

$$\{2(\frac{\delta_2\rho}{\bar{\rho}} - (\frac{\delta_1\rho}{\bar{\rho}})^2)\left(\frac{\delta_2 P}{\bar{P}}(\xi,\xi^*) + \frac{\delta_2 P}{\bar{P}}(\xi^*,\xi) + \frac{\delta_2 P}{\bar{P}}(\xi,\xi)\right)$$

$$+ \frac{\delta_1\rho}{\bar{\rho}}\left(\frac{\delta_3 P}{\bar{P}}(\xi,\xi,\xi^*) + \frac{\delta_3 P}{\bar{P}}(\xi,\xi^*,\xi) + \frac{\delta_3 P}{\bar{P}}(\xi^*,\xi,\xi)\right)\}$$

$$+ \frac{1}{2\sigma J}\int_M dm\; (2\sigma^2 + \sigma^*)\bar{r}^2\xi_r^4 \tag{B7}$$

The second and third order variations of pressure are given by (A6-A8). The symmetrization then leads to:

$$\frac{\delta_2 P}{\bar{P}}(\xi,\xi^*) + \frac{\delta_2 P}{\bar{P}}(\xi^*,\xi) + \frac{\delta_2 P}{\bar{P}}(\xi,\xi) =$$

$$3\left(\Gamma_1\frac{\delta_2\rho}{\bar{\rho}} + \frac{1}{2}\alpha_1(\frac{\delta_1\rho}{\bar{\rho}})^2\right) + [\frac{1}{2}\alpha_3(2|\xi_s|^2 + \xi_s^2) + \alpha_2\frac{\delta_1\rho}{\bar{\rho}}(2\xi_s + \xi_s^*)] \tag{B8}$$

$$\frac{\delta_3 P}{\bar{P}}(\xi,\xi,\xi^*) + \frac{\delta_3 P}{\bar{P}}(\xi,\xi^*,\xi) + \frac{\delta_3 P}{\bar{P}}(\xi^*,\xi,\xi) =$$

$$3\left(\Gamma_1\frac{\delta_3\rho}{\bar{\rho}} + \alpha_1\frac{\delta_1\rho}{\rho}\frac{\delta_2\rho}{\bar{\rho}} + \frac{1}{6}\alpha_4(\frac{\delta_1\rho}{\bar{\rho}})^3\right)$$

$$+ [(\alpha_2\frac{\delta_2\rho}{\bar{\rho}} + \frac{1}{2}\alpha_5(\frac{\delta_1\rho}{\bar{\rho}})^2)(\xi_s^* + 2\xi_s) + \frac{1}{2}\alpha_6\frac{\delta_1\rho}{\bar{\rho}}(\xi_s^2 + 2|\xi_s|^2) + \frac{1}{2}\alpha_7\xi_s|\xi_s|^2] \tag{B9}$$

Insertion of these expressions in (B7) gives rise to:

$$Q^c = \frac{3}{2\sigma J}\int_M dm\; [-\frac{\bar{P}}{\bar{\rho}}(q_{ad} - i\frac{\kappa}{\Omega}q_2) + 1/3(2\sigma^2 + \sigma^*)\bar{r}^2\xi_r^4] \tag{B10}$$

with

$$q_{ad} = 2\Gamma_1\left(\frac{\delta_2\rho}{\bar{\rho}}\right)^2 + 2(\alpha_1 - \Gamma_1)\frac{\delta_2\rho}{\bar{\rho}}\left(\frac{\delta_1\rho}{\bar{\rho}}\right)^2 + (\frac{1}{6}\alpha_4 - \alpha_1)\left(\frac{\delta_1\rho}{\bar{\rho}}\right)^4$$

$$+ \Gamma_1\frac{\delta_1\rho}{\bar{\rho}}\frac{\delta_3\rho}{\bar{\rho}} \tag{B11}$$

$$q_2 = d_0\frac{\delta_1\rho}{\bar{\rho}}\frac{\zeta}{\kappa} + 2(\frac{\kappa}{\Omega})^2 d_1(\frac{\zeta}{\kappa})^2 + (\frac{\kappa}{\Omega})^2 d_2(\frac{\zeta}{\kappa})^3 = d_0\frac{\delta_1\rho}{\bar{\rho}}\frac{\zeta}{\kappa} + O(\frac{\kappa}{\Omega})^2 \tag{B12}$$

where we have defined $\zeta = \sigma\xi_s$ i.e. from (7c)

$$\zeta = \sigma\xi_s = -\frac{\bar{L}}{c_v\bar{T}}\frac{d\delta_1 L/\bar{L}}{dm} \tag{B13}$$

and

$$d_0 = 3\alpha_2\frac{\delta_2\rho}{\bar{\rho}} - \frac{1}{2}(4\alpha_2 - \alpha_5)(\frac{\delta_1\rho}{\bar{\rho}})^2 \tag{B14}$$

$$d_1 = \frac{1}{2}(2\alpha_6 - \alpha_3)(\frac{\delta_1\rho}{\bar{\rho}})^2 + \alpha_3\frac{\delta_2\rho}{\bar{\rho}}; \quad d_2 = \frac{1}{6}\alpha_7\frac{\delta_1\rho}{\bar{\rho}} \tag{B15}$$

The α coefficients are defined in (A10-A14). We have neglected terms of order $O(\kappa/\Omega)^2$ and higher. This is valid as long as ζ/κ is much less than $(\kappa/\Omega)^{-2}$, which remains true far beyond the quasiadiabatic approximation break down.

Dismissing terms of order $O(\kappa/\Omega)^2$ in σ^{-1} and σ^2 in (B10), one finds that:

$$Q^c = -\kappa\Lambda_r + i\Omega\Lambda_i + O(\frac{\kappa^2}{\Omega}) \tag{B16}$$

with

$$\Lambda_r = \frac{3}{2}\int_M dm\ \xi^2\ \lambda_r(m); \qquad \Lambda_i = \frac{3}{2}\int_M dm\ \xi^2\ \lambda_i(m) \tag{B17}$$

$$\xi^2\ \lambda_r(m) \equiv \frac{1}{\Omega^2 J}\left(\frac{\bar{P}}{\bar{\rho}}(q_{ad} - d_0\frac{\delta_1\rho}{\bar{\rho}}\frac{\zeta}{\kappa}) + 1/3\Omega^2\bar{r}^2\xi_r^4\right) \tag{B18}$$

$$\xi^2\ \lambda_i(m) \equiv \frac{1}{\Omega^2 J}\left(\frac{\bar{P}}{\bar{\rho}}q_{ad} + 1/3\Omega^2\bar{r}^2\xi_r^4\right) \tag{B19}$$

Expressions A15-A17 for the density variations are next used in B11-B12. We now discuss the case of a marginally unstable fundamental mode and neglect the spatial derivatives of ξ_r in the density variations i.e. in B11-B12. Then q_{ad} reduces to

$$q_{ad} = 6c_0\ \xi_r^4 \quad \text{with} \quad c_0 = \frac{9}{4}(2\alpha_1 + \alpha_4) - \Gamma_1 \tag{B20}$$

Then, (B18-B19) becomes:

$$\xi^2\ \lambda_r(m) = \frac{1}{\Omega^2 J}\left(\frac{\bar{P}}{\bar{\rho}}(6c_0\xi_r^4 - d_0\frac{\delta_1\rho}{\bar{\rho}}\frac{\zeta}{\kappa}) + 1/3\Omega^2\bar{r}^2\xi_r^4\right) \tag{B21}$$

$$\xi^2\ \lambda_i(m) = \frac{1}{\Omega^2 J}\left(\frac{\bar{P}}{\bar{\rho}}6c_0 + 1/3\Omega^2\bar{r}^2\right)\xi_r^4 \tag{B22}$$

From (B14, B20, A10-A13), c_0 and d_0 are given by:

$$c_0 = \frac{9}{4}\Gamma_1[\Gamma_1(\Gamma_1 - 1) - \frac{4}{9} +$$
$$(3\Gamma_1 - 1)\left(\frac{\partial\Gamma_1}{\partial LogP}\right)_s + \left(\frac{\partial\Gamma_1}{\partial LogP}\right)_s^2 + \Gamma_1\left(\frac{\partial^2\Gamma_1}{\partial LogP^2}\right)_s] \tag{B23}$$

$$d_0 = \frac{9}{2}\alpha_5\xi_r^2 \tag{B24}$$

We finally end up with:

$$\lambda_r(m) = (9x + 1/3)\Phi_1(m) - 9y\Phi_2(m) \tag{B25}$$

$$\lambda_i(m) = (9x + 1/3)\Phi_1(m) \tag{B26}$$

where we have defined:

$$x = (\frac{c_s}{\Omega\bar{r}})^2 \frac{3}{2}\{\Gamma_1(\Gamma_1 - 1) - \frac{4}{9} + (3\Gamma_1 - 1)\left(\frac{\partial\Gamma_1}{\partial LogP}\right)_s + \left(\frac{\partial\Gamma_1}{\partial LogP}\right)_s^2$$

$$+ \Gamma_1\left(\frac{\partial^2\Gamma_1}{\partial LogP^2}\right)_s\} \qquad (B27)$$

$$y = \{\Gamma_3(\Gamma_3 - 1) + (3\Gamma_3 - 2)\left(\frac{\partial\Gamma_3}{\partial LogT}\right)_s + \left(\frac{\partial\Gamma_3}{\partial LogT}\right)_s^2$$

$$+ (\Gamma_3 - 1)\left(\frac{\partial^2\Gamma_3}{\partial LogT^2}\right)_s\} \qquad (B28)$$

$$\Phi_1(m) = \frac{\bar{r}^2\xi_r^2}{\int_M dm\ \bar{r}^2\xi_r^2} > 0; \qquad \Phi_2(m) = \frac{k(m)}{\int_M dm\ k(m)} \qquad (B29)$$

$$k(m) = -\frac{\bar{L}}{2J\Omega^2}(\Gamma_3 - 1)\frac{\delta_1\rho}{\bar{\rho}}\frac{d\delta_1 L/\bar{L}}{dm} \qquad (B30)$$

and $c_s = (\Gamma_1\bar{P}/\bar{\rho})^{1/2}$ is the sound speed.

Turbulence, Fractals, and the Solar Granulation

P.N. Brandt [1], R. Greimel [2], E. Guenther [3], W. Mattig [1]

[1]Kiepenheuer-Institut für Sonnenphysik, Schöneckstr. 6, D–W–7800 Freiburg–im–Breisgau, Germany

[2]Institut für Astronomie, Universitätsplatz 5, A–8010 Graz, Austria

[3]Max-Planck-Institut für Astronomie, Königstuhl 17, D–W–6900 Heidelberg 1, Germany

Abstract: We give a brief, mostly qualitative introduction into the topics of convection and turbulence, and their description, mainly referring to laboratory experiments. Following Mandelbrot (1967), the concept of the fractal dimension is introduced and some earlier results of measurements of the fractal dimension in laboratory turbulence are discussed. Next, we address the question whether hints of turbulence have been observed in the solar photosphere, and describe three independent methods of determining the fractal dimension of the solar granulation: the area-perimeter relation, the line and the plane intersection method. An analysis of a set of high resolution granulation photographs taken with the balloon-borne 'Spektro-Stratoskop' telescope yields a fractal dimension of $d \approx 1.9$ for granules of diameters > 1.32 arcsec and of $d \approx 1.3$ for smaller granules, analysing the area-perimeter relation of approx. 40 000 granules. At first sight we seem to confirm the results obtained by Roudier and Muller (1986), Darvann and Kusoffsky (1989), and by Karpinsky (1990), who claim to see a splitting of the granulation into two regimes of different fractal dimension. However, a more detailed investigation of the analysis technique reveals that: i) there exists a smooth transition of the fractal dimension from small to large granules, and ii) the fractal dimension of small granules seems to be dominated by the finite resolution, and therefore no positive statement concerning the turbulent origin of small granules seems possible with the present technique. The fractal dimensions determined from the same material with the other two methods (i.e. the line and the plane intersection method), applied to both the intensity pattern itself and the lane map, all range between 1.88 and 1.97.

1. Introduction

The study of turbulent phenomena has been a demanding task in many fields of physics and engineering. Especially in astrophysics, the description of turbulent energy transport and dissipation processes in stellar plasmas is of key importance for the understanding of the energy balance of stars. The photosphere of the Sun is one of the few places in astrophysics where *turbulent motion* could, in principle, be *directly* observed. Two achievements of the last $1\frac{1}{2}$ decades seemed to make it worthwhile to attack the ambitious task to investigate whether these topmost layers of the Sun, accessible to visual observation, indeed are in turbulent motion. The first achievement is the high spatial resolution of 0.25 to 0.3 arcsec (180 to 220 km on the Sun), which can now be attained from excellent sites with carefully designed solar telescopes of apertures in the range from 50 to 60 cm and with advanced observation techniques. The second achievement is the development of new ways to describe and analyze chaotic systems, i.e. the concept of fractals.

After some introductory remarks on the basic properties of convection and turbulence, and a brief presentation of the concept of fractals, we present in the following some results of investigations of turbulent flows, both in laboratory experiments and in meteorology. A separate chapter is dedicated to the discussion of recent results concerning observations of the characteristics of turbulence in the solar photosphere, as derived from measurements of velocities as well as from analyses of the structure of the granulation. Finally, we present new results of an investigation of granular structure using a series of high resolution observations and discuss their relevance with respect to the question, whether the existence of turbulent motions in the upper layers of the solar photosphere can be proved or disproved.

2. Convection and Turbulence

Convection is the dominant mechanism of energy transport in the envelopes of many stars, as for example the Sun. The flow of the free convection is driven by buoyancy forces which are induced by a temperature gradient between the lower and upper boundaries of the plasma in a gravitational field. The condition for the onset of convection in a star is given by the Schwarzschild criterion (see e.g. Stix 1989): if the adiabatic temperature gradient is smaller than the one given by radiative equilibrium, then the bulk of the energy transport is taken over by convective processes. Convection has been studied extensively, both in theoretical treatments and in laboratory experiments (for an introduction into the topics of convection and turbulence see e.g. Bradshaw and Woods 1976; Tritton 1977; Bray et al. 1984; Rüdiger 1989; Spruit et al. 1990).

In laboratory experiments at extremely low flow velocities a regular pattern of flow cells is found, as is well known from the Bénard experiment. At very high velocities one recognizes the seemingly irregular and unsteady *turbulent* motion of the fluid, which at first sight seems to escape a quantitative description. This latter type of convection is of main importance for astrophysical applications, especially for stellar convection zones. Deeper insight into the turbulence phenomenon seems thus desirable, and one may find out that turbulence is somewhat better accessible, at least to a formal description, than expected. But, *what is turbulence*? In a first approach, turbulence could be defined somewhat vaguely as an irregular and unsteady motion of the fluid.

Imagine a flow in a pipe. If the flow velocity is very low, one observes the well ordered laminar Poiseuille flow. At higher flow velocities, the sudden occurrence of so-called turbulent plugs is observed in the flow. In these plugs the velocity shows rapid irregular fluctuations, and we denote this as turbulence. Thus laminar and turbulent flows are observed in the pipe at the same time, which is called 'intermittency'. When the velocity is increased further, the fraction of turbulent motion will increase, and finally the flow in the pipe becomes totally turbulent. In order to describe the transition from laminar to turbulent flow quantitatively, the dimensionless Reynolds number is introduced:

$$Re = \frac{\rho \cdot v \cdot l}{\mu}\,, \tag{1}$$

where v is the velocity, ρ the mass density, μ the viscosity of the fluid, and l the diameter of the pipe.

From various experiments (see for example Tritton 1977) one concludes that the transition to turbulence occurs at Reynolds numbers between 2×10^3 and 1×10^5, depending strongly on the finer details of the experiment (Tritton 1977). At this point, we may ask ourselves, what the order of the Reynolds number in the solar convection zone is? Assuming l to be about 100 km, which is the pressure scale height at photospheric levels, Unsöld (1955) estimated a value of the order of 10^9. We thus conclude that turbulence is indeed important, although stellar convection zones are not pipes.

Fully developed turbulence was often thought of as a hierarchy of eddies of different scales. This should be the result of a loss of stability of the largest eddies (Frisch et al. 1978). Kolmogorov (1941) and others introduced the view that the energy is injected in the large scale eddies, and then cascades down to smaller scales, where the energy is dissipated. If we define E(k) as the kinetic energy per unit mass and unit wavenumber, ϵ the rate of energy dissipation, then Kolmogorov's law for the energy spectrum is:

$$E(k) \sim \epsilon^{2/3} k^{-5/3}\,. \tag{2}$$

This law can be derived by dimensional analysis, but Frisch et al. (1978) derived it using dynamical arguments. In deriving the Kolmogorov law one assumes that at a given moment there exists only one eddy size, and some time later this size has decreased. In order to save the concept of *simultaneous existence* of large and small eddies, Frisch et al. (1978) derived the relation

$$E(k) \sim \epsilon^{2/3} k^{-5/3} (kl)^{-(1-D/3)}\,, \tag{3}$$

where l is the scale of the largest eddies, and D the self-similar dimension. This leads naturally to the introduction of the self-similar dimension D, which is a special case of Mandelbrot's fractal dimension D (Mandelbrot 1975).

3. Fractal Dimension and Self-Similarity

In the following, we try to give a brief, and necessarily incomplete, description of the concepts of the fractal dimension and self-similarity. For a detailed introduction into the the topic of fractals the interested reader is referred to the classical textbook by Mandelbrot (1982), as well as the books by Peitgen and Richter (1986) and by Feder (1989). The analysis of fractal phenomena in a variety of fields in physics is treated in the proceedings volume edited by Aharony and Feder (1990).

A nice way to introduce the concept of fractals was used by Mandelbrot (1967) asking: How long is the coast of Britain? Obviously, the length of a rugged, irregular and winding coastline is not a well defined thing, depending strongly on the scale at which one looks at it and on the measuring procedure. For example, using dividers set to a prescribed opening δ, or alternatively a yardstick of the same length, and 'walking' them along a coastline map, one finds a certain number of steps necessary to cover the whole perimeter of the island. The number of steps N multiplied by δ is an approximate measure of the length $L(\delta)$. However, setting the divider opening to a smaller value, one will obtain a larger value of the length $L(\delta)$, because more little inlets and peninsulae are included in the measurement. Using smaller and smaller divider spacings and maps of finer and finer scales, $L(\delta)$ seems to grow to infinity, which suggests that no meaningful result at all can be obtained. A way out of this dilemma is to plot the length of the perimeter $L(\delta)$ as a function of δ on a log – log scale. It turns out the relation can be represented by a straight line described by

$$L(\delta) \sim \delta^{1-D} , \tag{4}$$

$1 - D$ being the slope of the regression line. Typical values of D between 1.3 and 1.5 were found for real coastlines. In a generalization of the term dimension, Mandelbrot (1967) coined the term 'fractal dimension' for this D, since it can attain non-integer values between 1 and 2. For an 'ordinary' curve, like the perimeter of a circle, the procedure would result in a constant $L(\delta)$ for values of δ below a certain limit, thus leading to a slope of zero and $D = 1$, and we come back to the simple fact that it is a one-dimensional object in the 'old' sense. The famous Koch curve, whose perimeter consists of triangular shaped structures of finer and finer scale, has a fractal dimension $D = 1.26$. In general terms, the fractal dimension enables us to describe the morphological complexity of boundaries, and very similar generalized definitions had been worked out for two-dimensional and three-dimensional objects. Three specific methods of determination of the fractal dimension are described in Sect. 6.

An important feature of fractals is the so-called self-similarity: if a small subsection of the object is presented at a magnified scale it shows the same structuring again, as is nicely demonstrated by Peitgen and Richter (1986). This makes the concept of fractals suitable for the description of turbulent processes, where a hierarchy of eddies of different sizes is believed to exist.

4. Fractal Dimension of Turbulence

Mandelbrot (1975) studied the geometry of homogeneous Kolmogorov $k^{-5/3}$ turbulence and concluded that iso-surfaces (e.g. surfaces of constant temperature or concentration) can be characterized by a fractal dimension of $3-\frac{1}{3}$ in three dimensions, whereas two and one-dimensional cuts have a fractal dimension of $2-\frac{1}{3}$ and $1-\frac{1}{3}$, respectively. Mandelbrot (1977) goes even as far, as to say that a definition of fully developed turbulence must include the fractal dimension. Sreenivasan and Meneveau (1986) tried to solve the question experimentally, whether constant-property surfaces (e.g. iso-velocity, iso-concentration, or iso-bars, and so on) are fractals in fully developed turbulence, and if so what their fractal dimension is. They concluded that their work fell short of definitely proving that turbulence is a fractal, because of the limitations of the techniques employed. The authors believed that turbulence is perhaps a collection of a number of fractals each of which is slightly different, and there is not a one-to-one correspondence between geometry and dynamics. But anyway, if we take their measurements seriously, the values for the fractal dimension are:

$D = 2.6$ for temperature dissipation field,
$D = 2.7$ for iso-dissipation lines, and
$D = 2.4$ for the border lines between turbulent and non-turbulent areas [14].

A fractal dimension of $2-\frac{1}{3}$ for atmospherical iso-thermal cuts and $2-\frac{2}{3}$ for iso-bar cuts was proposed by Lovejoy (1982) using Mandelbrot's (1975) results and theoretical arguments. From analyzing clouds of sizes between 1 and 1000 km in digitized satellite images he indeed found a fractal dimension of 1.34, which would be in agreement with iso-bar cuts. In a critical analysis of this result, Sreenivasan and Meneveau (1986) argued that the satellite images are not cuts in the atmosphere, but projections. In order to evaluate the difference of D measured in a projection of a layer of finite thickness instead of an iso-cut, Sreenivasan and Meneveau (1986) increased the thickness of the observed layer in one of their experiments and found an *increase* of the fractal dimension. They then argued that the real fractal dimension of clouds must be lower than measured by Lovejoy (1982). As a confirmation they cited the results of Carter et al. (1986) who found a fractal dimension of 1.16.

It is important to note that different methods were used to evaluate the fractal dimension in the investigations mentioned above. These methods, and their drawbacks, are so important for the discussion of the results that we will discuss them in detail below. We notice that Lovejoy (1982) used the area-perimeter method,

[14]Note that this D is a measured value, which is not strictly measured according to the mathematical definition of the self-similar dimension.

whereas Sreenivasan and Meneveau (1986) used the line and the plane intersection method.

5. Turbulence in the Solar Photosphere

The existence of a convection zone in the Sun is a well established fact. But, this zone is *not directly observable*. There are two important different ways to study some of the physics of the solar convection zone: helioseismology (cf. Deubner and Gough 1984) and observation of the overshoot layers. In the following, we concentrate on the latter only. Convective elements are overshooting into the convectively stable layers immediately above, and can be observed as the well known granulation pattern (for a presentation of the theoretical backgrounds of solar convection and overshoot cf. Bray et al. 1984; Stix 1990). The penetration is about 150 km (Nesis et al. 1988; Komm et al. 1990). From the observation of granules we try to find out something about the convection, but we have to remember that we are not observing the convection zone. The physics of solar and stellar convection zones are summed up e.g. by Zahn (1987).

Has the existence of turbulence in the upper layers of the solar photosphere been observed? There seems to be contradicting evidence concerning this topic. Let us first discuss *velocity measurements*. On the one hand Muller et al. (see Zahn 1987) and Komm et al. (1990) found Kolmogorov's −5/3–law in power spectra of solar photospheric velocities for scales from approx. 3 arcsec (2 Mm) down to the resolution limit of their observations (approx. 0.4 Mm). From this fact they inferred on the existence of turbulence in this regime. On the other hand, Deubner (1988) was unable to detect any hints of a critical scale around 1.4 arcsec, as reported by Roudier and Muller (1986, see below) and finds temporal and spatial phases to remain stable down to 0.8 arcsec, thus denying hints of turbulence. Wiehr and Kneer (1988) as well as Nesis and Mattig (1989) observed the correlation of intensity and velocity to break down at scales below 1.5 arcsec, and interpret this as rapid breaking of the convective overshoot or the onset of turbulence at these scales.

From an analysis of the structure of the *intensity pattern* of the granulation Roudier and Muller (1986) found that it is split up into two regimes: below granule diameters of 1.37 arcsec a fractal dimension d=1.25, above that diameter d=2.15 was found. From this fact they infer on the turbulent origin of the smaller granules. Very similar results were obtained by Darvann and Kusoffsky (1989) and by Karpinsky (1990). In the following, after a brief presentation of the methods of fractal dimension analysis, we present some results derived from a series of high resolution observations, and then come back to the discussion of the relevance of the results cited above.

6. Methods of Determining the Fractal Dimension of Solar Granulation

The solar granulation is a phenomenon in three dimensions. But it is not possible to measure the true three-dimensional distribution of e.g. the temperature, pressure or velocity. On the other hand it is possible to determine a two-dimensional distribution of the intensity from white light pictures. There are several methods available to calculate the fractal dimension of a two-dimensional data set. Two of these will be discussed below. In order to be able to relate results from those to the original three-dimensional object, we have to think about lower dimensional subspaces (see e.g. Sreenivasan and Meneveau 1986). In analogy to Euclidian geometry we may expect that the fractal dimension D_2 of the border of a two-dimensional cut (plane intersection) is one less than the dimension of the surface of the three-dimensional object (D_3):

$$D_2 = D_3 - 1 \; .$$

The same should be true for one-dimensional cuts (line intersections) leading to

$$D_1 = D_3 - 2 \; .$$

6.1 Area-Perimeter Relation

One method to determine a fractal dimension is the area-perimeter relation. The dimension calculated is the dimension of the perimeter of the fractal. Suppose that we have similar islands of different size and therefore different area. Now imagine that we measure all those differently sized islands with the same yardstick of length δ. Referred to the area of the islands the yardstick then has relative lengths, which vary like

$$\delta_i \sim A_i^{-1/2} \; . \tag{5}$$

The index i runs over the number of islands. We may derive the length of the perimeter of the ith island by inserting (5) into (4)

$$L_i(\delta_i) \sim \delta_i^{1-d} \sim A_i^{-1/2+d/2} \sim A_i^{d/2} . \tag{6}$$

Constant factors have been neglected in the last two equations. In order to distinguish the fractal dimensions derived with different methods, we denote the fractal dimension determined from the area-perimeter relation by d. The fractal dimension can then be determined by calculating the slope of the area vs. perimeter plot:

$$d = 2 \times \frac{\Delta \log(L)}{\Delta \log(A)} \; . \tag{7}$$

A correct equation for the area-perimeter relation is derived e.g. by Feder (1989, p. 200 ff):

$$L(\delta) = N_\lambda \lambda^d \delta^{1-d} \sqrt{A(\delta)}^{\,d} , \tag{8}$$

where λ is an arbitrary but small number, and N_λ is the number of segments with length δ of the approximating polygon. This equation holds for any yardstick δ small enough to measure the small islands accurately.

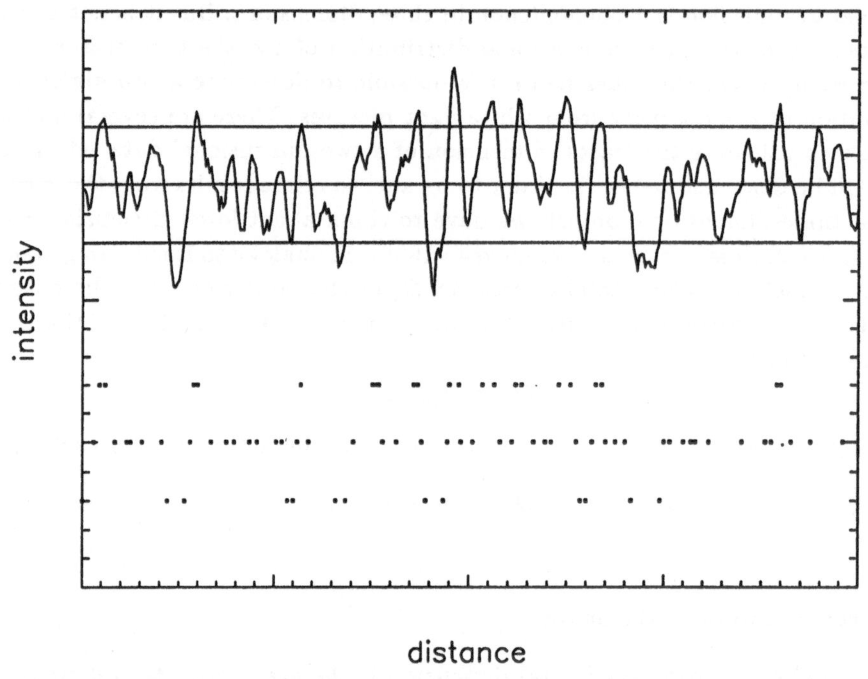

Fig. 1. The principle of the line intersection method. By intersecting the measured function with a line at a specific level, one generates a Cantor set with a dimension D_1 lower than 1.

6.2 Line Intersection Method

The principle of the line intersection method is shown in Figs. 1 and 2 (cf. also Sreenivasan and Meneveau 1986). By intersecting a measured function with a line of constant value one generates a Cantor set of points. The next step is to count the number $N(\delta)$ of the segments needed to cover the set for different values of the yardstick length δ. In other words, one places yardsticks of a given length next to each other covering the total interval and counts the number of yardsticks that contain at least one intersection point; this procedure is repeated for different values of δ. The dimension is then calculated from the slope of the $\log N(\delta)$ vs. $\log \delta$ plot (cf. Fig. 2), and is denoted by D_1. Obviously, the most dense set of points will be produced by a cut near the level of the mean intensity. This is shown in Fig. 1 for cuts through the solar granulation intensity distribution.

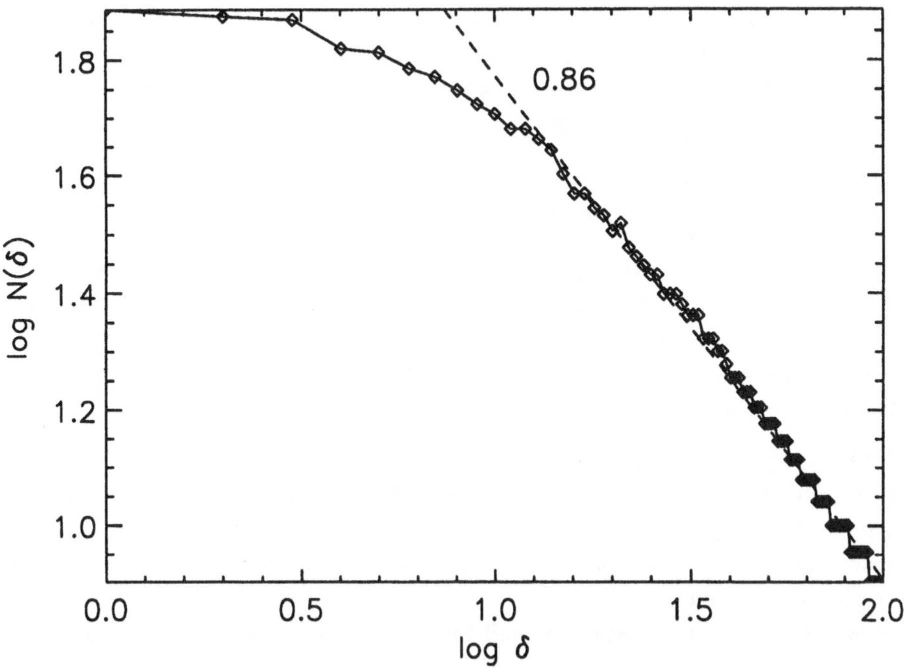

Fig. 2. Number of segments $N(\delta)$ that contain at least one intersection point as a function of the yardstick length δ in a log – log presentation. The fractal dimension D_1 is calculated from the slope of the linear fit.

6.3 Plane Intersection Method

The plane intersection method is the logical extension of the line intersection method into two dimensions. The intensity distribution is cut at a specific level, thus generating closed contour lines of iso-intensity. Now, instead of covering the interval with yardsticks δ placed next to each other, one covers a plane with squares of side-length δ also placed next to each other. As in Sect. 6.2 the next step is to vary δ and count the number of squares that contain at least one contour point. The slope of the resulting linear fit to the $\log N(\delta)$ vs. $\log \delta$ plot is the fractal dimension of the plane intersection method, denoted by D_2.

7. Observational Material and Methods of Analysis

7.1 Observational Material

For the present analysis observing material from the balloon-borne solar telescope 'Spektro-Stratoskop' (aperture 32 cm; cf. Mehltretter 1978) was used. The observations were done on May 17, 1975 during a flight at an altitude of 27 km. The instrument package contained a spectrograph and a photographic white light slit-jaw camera (5560 ± 125 Å). The advantage of these white light pictures is their photometric calibration and nearly no degradation due to seeing. A set of 13 frames with time lags of approx. 3 minutes was selected for further analysis. The spatial resolution of the pictures is approximately 0.4 arcsec, their scale 2.98 arcsec/mm, and the total area covered on the Sun is 72 × 108 arcsec^2.

7.2 Methods of Analysis and Data Correction

The photographs were microphotometered using a quadratic aperture of 55 × 55 μm (this corresponds to 0.15 × 0.15 arcsec on the Sun) and a step size of 50 μm thus leading to arrays of approx. sizes of 450 × 750 pixels per picture.

A close inspection of the pictures revealed essentially three types of defects which had to be corrected:

- intensity gradients up to 9 % over the whole picture;
- the defocused spectrograph slit section of the images had to be restored (because of the defocused appearance of the slit on the white light pictures, the intensity information underneath the slit was not lost but lowered, and thus could be restored);
- scratches and 'dust' were removed by a line and point mask procedure (Gonzales and Wintz 1987).

7.3 Lane Finding

In principle, there is no clear definition of a granule from an intensity picture alone (i.e. without having velocity information) and, therefore, many descriptions are just operational definitions by the authors (Collados et al. 1986; Roudier and Muller 1986; Title et al. 1987; Darvann and Kusoffsky 1989).

Our algorithm is similar to the one used by Title et al. (1987), called 'lane finding'. This procedure essentially determines the intergranular lanes and dark structures within granules and delivers 'dual' lane maps consisting only of 'zeros' (representing lane pixels) and 'ones' (granule pixels). A section of such a lane map is shown in Fig. 3 together with the corresponding granule pattern. As a first step, the whole picture is searched for *local minima*. A local minimum is defined as an intensity minimum in at least one direction, either horizontal, or vertical, or one of the two diagonals. Each pixel in the image is checked for being a local minimum and set as a lane pixel if this is the case. The intergranular lanes are quite narrow structures at the resolution limit of the telescope. One would expect the intensity of the intergranular lanes to be lower than that of the neighbouring pixels of granules. But not all lane pixels are *local* minima. Therefore a second condition for determining lane pixels has to be introduced. This is accomplished by forcing all pixels in the image with an intensity $I < m - 2\sigma$ to be lane pixels (m denotes the mean and σ the standard deviation of the intensity pattern). On the other hand, pixels with $I > m + 0.7\sigma$ are set as granule pixels. The second condition prevents the splitting up of granules to some extent. The lane map produced by the algorithm up to now still shows gaps in the lanes which have to be filled.

Therefore, the lane map is searched for *one* pixel gaps between the lane pixels. These gaps are closed, i.e. the gap pixels are set as lane pixels, if $\Delta I < \sigma$ with ΔI being the intensity difference of the neighbouring pixels of the gap. This last procedure is repeated until no more lane pixels have to be added to the lane map.

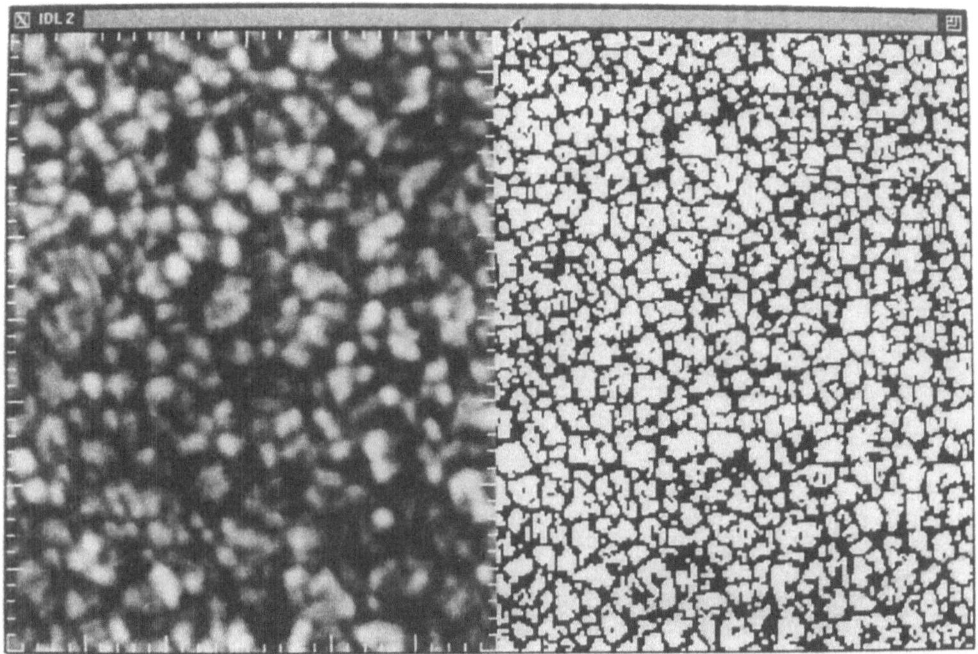

Fig. 3. Left: Solar granulation picture taken with the balloon-borne solar telescope 'Spektro-Stratoskop'; field size 30 by 37.5 arcsec^2. Tickmarks are 1 arcsec. Right: Same field converted into a lane map by the lane finding algorithm described in the text.

The lane maps are then scanned for granules by a recursive algorithm. The first step is to search for a granule pixel. The pixels are then marked as used, and the neighbouring pixels in the horizontal and the vertical direction are checked for being granular pixels, but *no* check is done in the two diagonal directions. If one neighbouring granular pixel does exist, then the above stated procedure is repeated for that pixel again. This is done until all pixels in the granule have been found, i.e. if their horizontal or vertical neighbour pixels are either lane pixels or marked as used.

8. Results and Discussion

8.1 Fractal Dimension from the Area-Perimeter Relation

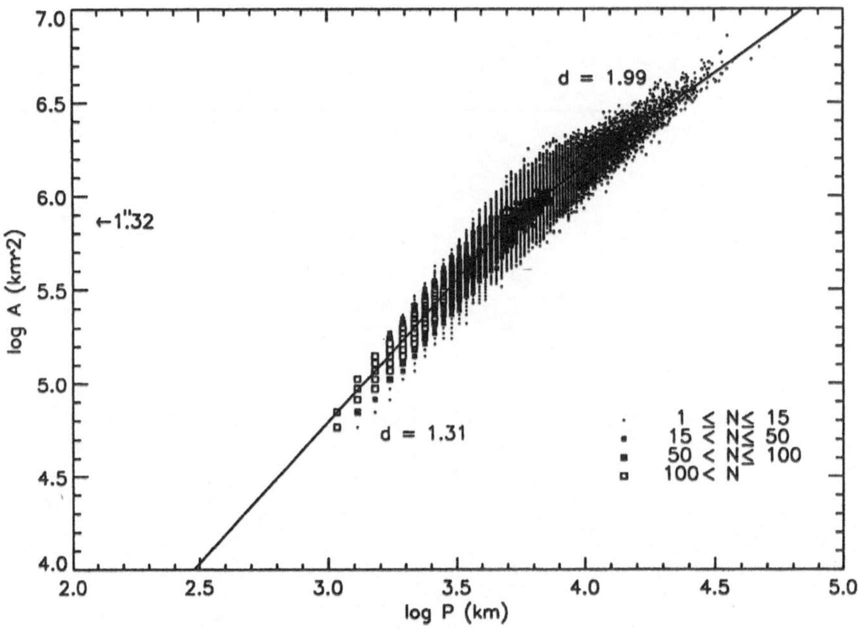

Fig. 4. The area-perimeter relation for 42 742 granules. The different symbols indicate the number N of granules with the same area and perimeter. The two straight lines represent linear fits to the data. The change in the fractal dimension (d) occurs at a granular diameter of 1.32 arcsec.

For the analysis of the fractal dimension, 13 lane maps with a time difference of approximately 3 minutes were used. Because of the mean granular lifetime, which is around 10 minutes, the granulation pattern has changed significantly from one picture to the next. Approximately 50 000 granules were identified on the lane maps. Their area and perimeter was determined as well. For further calculations only granules with areas larger than 4 pixels, i.e. $0.3 \times 0.3\,\mathrm{arcsec}^2$ or $5 \times 10^4\,\mathrm{km}^2$, were used. The reason for choosing this lower limit was that no information is to be expected at scales below the resolution limit of the observation (≈0.4 arcsec). This left 42 742 granules usable for the fractal dimension analysis using the area-perimeter relation (cf. Sect. 6.1). The resulting area-perimeter plot is shown in Fig. 4. At first sight our results seem to reproduce those obtained by Roudier and Muller (1986), Darvann and Kusoffsky (1989), and by Karpinsky (1990), i.e. the existence of two different regimes, one with a fractal dimension around 1.3 and another with $d \approx 2$, with the change of d occurring somewhere in the range of 1.3 to 1.8 arcsec for the granular diameter.

However, a detailed inspection reveals some limitations of the method, and therefore sheds some doubt onto its results and the conclusions drawn. These limitations mainly arise from:

- the finite pixel size,
- the limited range of size of the granules under study, and
- the resolution limit of solar observations.

Because of the finite pixel size, a minimum and a maximum possible perimeter, denoted by P_{min} and P_{max}, do exist for a given area:

$$P_{max}(A) = 2 \times A + 2 \tag{10}$$

$$P_{min}(A) = 4 \times int(\sqrt{A}) + \begin{cases} 0 \; for \; R = 0 \\ 2 \; for \; 0 < R \leq int(\sqrt{A}) \\ 4 \; else \end{cases} \tag{11}$$

with

$$R = A - [int(\sqrt{A})]^2 \,. \tag{12}$$

The area A has to be given in *pixels* in these equations. These two borderlines for the perimeter are shown in Fig. 5 together with the evaluated area-perimeter distribution. The two dashed lines show the slopes for the fractal dimensions of 1 and 2.

What do granules with P_{min} and P_{max} look like? For example a granule with P_{max} may show all pixels lined up in a row. Arranging the pixels to form granules with P_{min} is a bit more sophisticated. For area values (in pixels) whose square root is an integer value, the most dense package is achieved by placing them as a square. For any other area values the structures having the lowest perimeter are as close to squares as possible. It has to be mentioned that for a given area many different looking structures may reach the values of either P_{min} or P_{max}. But the structures described above are of such kind, that they *always* will lead to the minimum or maximum perimeter value.

In the following, we present some arguments for the fact that a fractal dimension of the order of 1.3 must be found in the *small scale regime* for *technical* reasons, independent of the real granulation structure:

a) It is obvious that all granules must lie between P_{min} and P_{max} (cf. Fig. 5). For one and the same area A, only a few values of $\log P$ can be attained.
b) It is important to note that single points in the lower left sections of Figs. 4 and 5 represent many granules.
c) Fig. 5 shows that the values of P_{min} and P_{max} are indeed attained. In the range $\log A \leq 5.6$ many granules show the value of P_{min}, and are clustered nearby. Just the opposite is true for P_{max}.
d) Due to the statistical weight of the large number of small granules, any linear fit is strongly biased by the behaviour of the smallest granules. This influence even propagates into the medium scales.

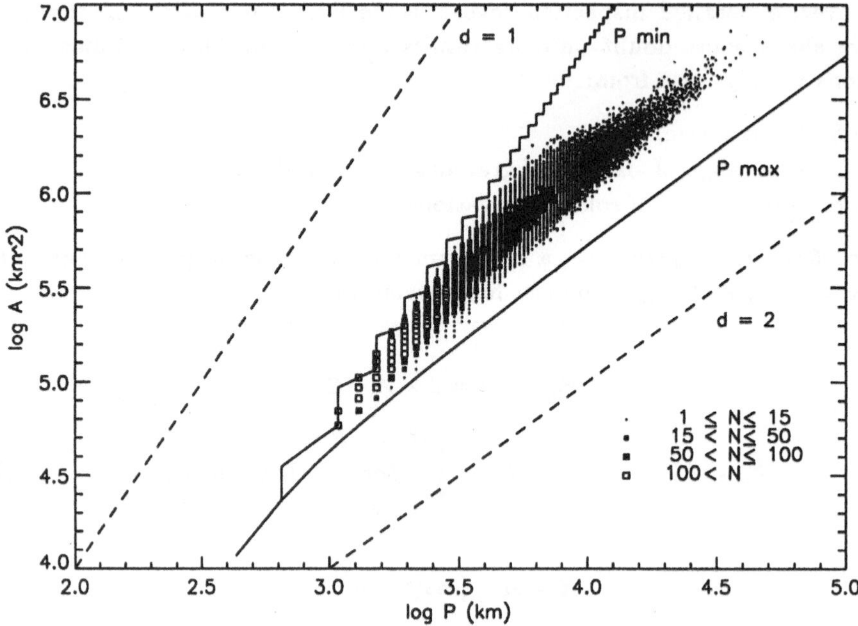

Fig. 5. The same as Fig. 4, with the curves for $P_{\min}$ and $P_{\max}$ added. The dashed lines show the slopes for fractal dimension d = 1 and d = 2, respectively. The different symbols indicate the number N of granules with the same area and perimeter.

The reasons for points a) and b) lie in the way of digitizing the photographs into pixels of finite size. Point c) might find an obvious explanation in the following: at its resolution limit, the telescope reimages a point source into the Airy disk. Also in a certain range above this limit, any small scale structure is filtered out by the convolution with the point spread function. Thus, *even if highly structured granules near the resolution limit would exist, they would be transformed into smooth and round patches.*

Looking at large size granules, the limited resolution and the pixel size is no longer a problem, but other problems come up. First of all, there is the limited number of granules of these sizes. Other problems come from the granulation definitions. Large granules may be produced by an incomplete splitting up of the granules. On the other hand, one has to take some care not to split up the granulation too much. This would lead to an even greater number of small granules.

Let us have a closer look at the determination of the fractal dimension despite all the problems mentioned above. In order to find out whether the area-perimeter relation indeed has a kink in the slope at intermediate sizes, we applied the following procedure. Instead of determining the fractal dimension for just two parts of the area-perimeter relation, we determined the slope of the linear fit in a small window of $\log P$. We then moved the window through the whole range of $\log P$,

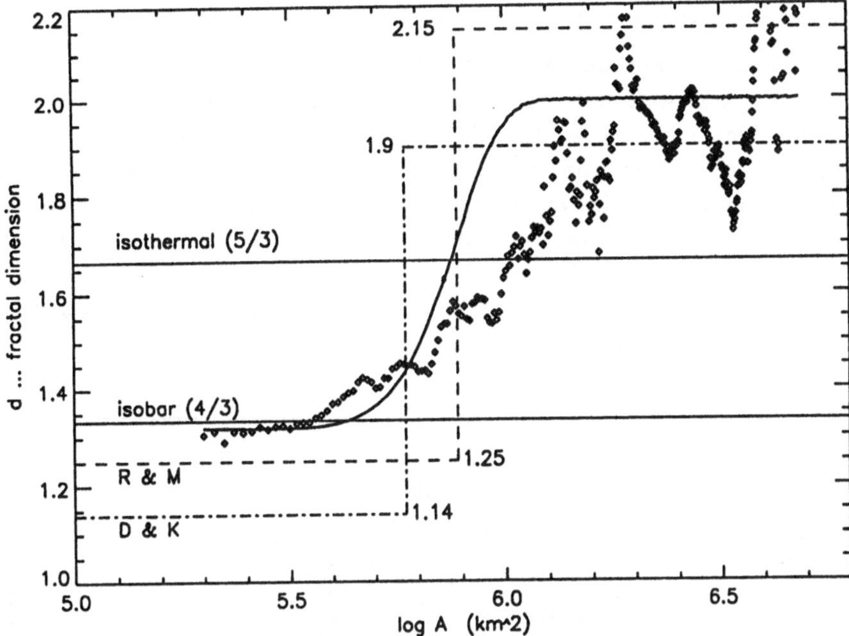

Fig. 6. The change in fractal dimension d, determined by the moving window method, for the data of Fig. 4 (◇). The lines at d = 4/3 and 5/3 are the fractal dimensions for iso-bars and iso-therms in the case of homogeneous, isotropic turbulence. The result for using the method on a sample distribution of two straight lines with a kink at a granular diameter of 1.32 arcsec, as seen in Fig. 4, is shown as the solid line. The dashed line, marked R&M, shows the results of Roudier and Muller(1986); the results of Darvann and Kusoffsky (1989) are shown by the dash-dotted line marked D&K.

each time calculating the slope of the linear fit. The result can be seen in Fig. 6 (diamonds). It is clearly seen that the fractal dimension does not jump from a low value to a high one, but instead *changes smoothly* over the whole range of sizes observed. For a test of the 'two regimes' hypothesis, we set all the granular perimeters of a given area to the corresponding values of the straight line fits seen in Fig. 4. Then our moving window method was used to determine the fractal dimension. As can be seen in Fig. 6, the change in dimension for the two-line distribution of the granules (solid line in Fig. 6) is quite sharp and occurs at the granulation diameter given in Fig. 4, which corresponds to log $A \approx 5.9$. This disproves the existence of two distinct regimes.

8.2 Fractal Dimension from the Line Intersection Method

The line intersection method can be applied to the intensity distribution of the granulation itself, as well as to the lane map. After choosing a cut level one determines the dimension D_1 as described in Sect. 6.2. The next step is to vary the cut level. Fig. 7 shows the results of this procedure applied to one frame of the Spektro-Stratoskop series of granulation photographs. Obviously, the fractal dimension obtained in this way varies in a wide range with its maximum value of $D_1 = 0.88$ (corresponding to $D_2 = 1.88$) near the mean intensity.

Applying the line intersection method to the lane map, on the other hand, yields a value of $D_1 = 0.97$ ($D_2 = 1.97$).

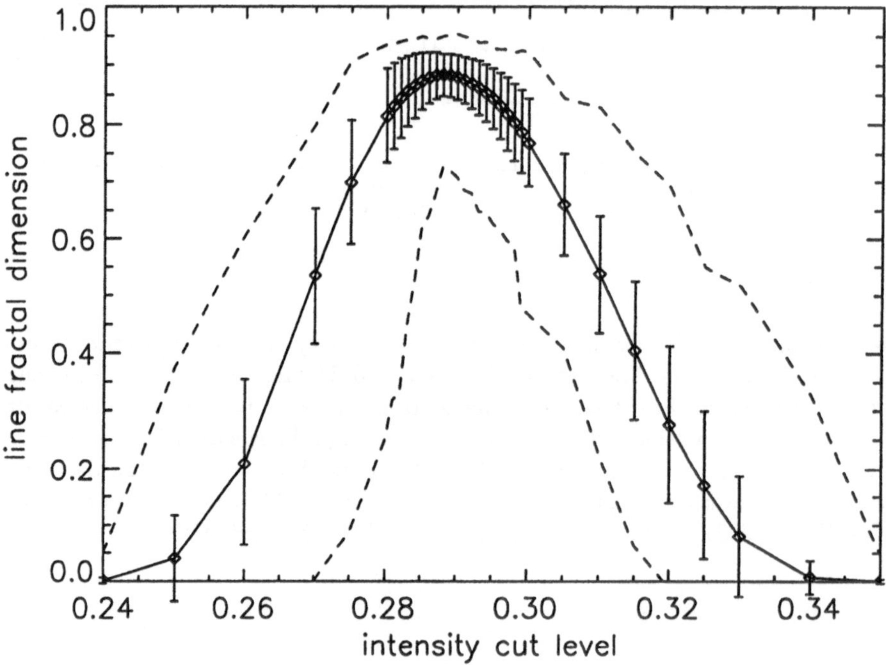

Fig. 7. Dependence of the line fractal dimension on the intensity cut level for one frame. The peak of the dimension occurs near the mean intensity of 0.29. The dashed lines indicate the minimum (below the solid curve) and the maximum values of the line fractal dimension that occurred in calculating 464 lines. The error bars extend to mean $\pm\ \sigma$.

8.3 Fractal Dimension from the Plane Intersection Method

The application of the plane intersection method is the same as in the case of the line intersection method. After choosing a cut level the procedure described in Sect. 6.3 is used. Thereafter the cut level is varied. The result is shown in Fig. 8 for one frame. After a sharp increase of the fractal dimension a stable value around $D_2 = 1.93$ is established, followed by a sharp decrease at high intensity values. Applying this method to the lane map yields a value of $D_2 = 1.93$.

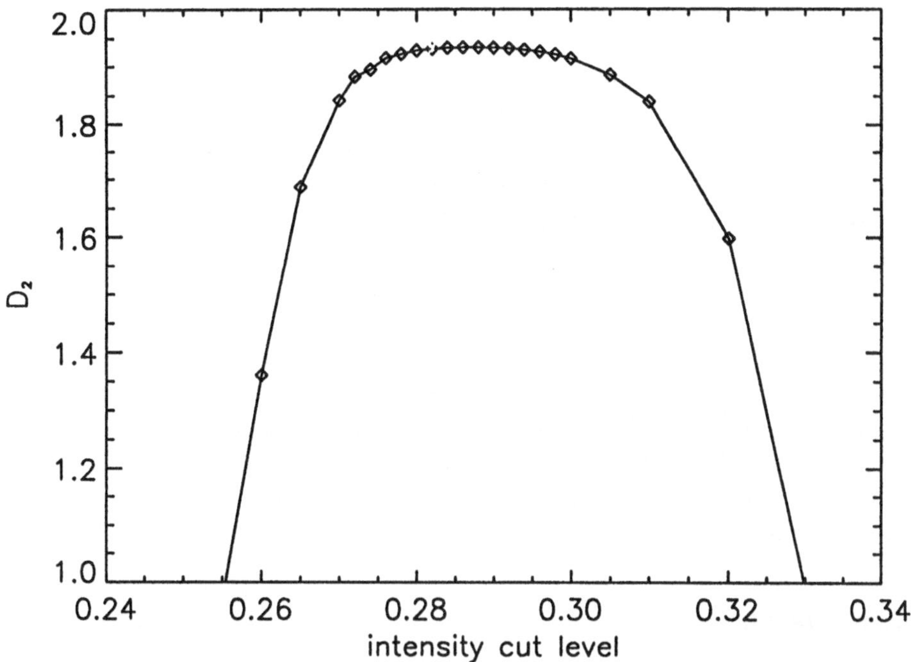

Fig. 8. Dependence of the fractal dimension D_2 on the intensity cut level, determined for one frame by the plane intersection method. Mean intensity is at 0.29 in relative units.

8.4 What is the 'True' Fractal Dimension of the Granulation?

Our above results were derived from white light pictures of the solar granulation. The contours of granules should, therefore, be closely related to iso-temperature surfaces, and a fractal dimension of $2 - \frac{1}{3} = 1.67$ is to be expected according to Mandelbrot (1977) if we are dealing with a turbulent phenomenon. It is not clear why Roudier and Muller (1986), Darvann and Kusoffsky (1989), and Karpinsky (1990) claim that their results in the range $1.14 \leq d \leq 1.25$ for smaller granules 'suggest that the smaller granules are of turbulent origin'. The values found by these authors indeed lie near 1.33; this is, however, the value expected for iso-bar cuts. To first approximation the intensity is given by the source function at optical depth $\tau = 1$ and in local thermodynamical equilibrium only depends on the temperature. Therefore, iso-intensity contours of the granulation should correspond to iso-thermal contours (rather than iso-bars), and 1.67 would be expected for the fractal dimension.

When comparing results from the analysis of the granular intensity distribution itself with those derived from the lane map, one has to keep in mind that the latter is strongly dependent on the somewhat arbitrary definition of what is 'a granule'. Nevertheless, the fractal dimensions determined for the large granules ($d \approx 1.9$) as derived from the area-perimeter relation and the four values obtained from the line and the plane intersection methods as given in Sects. 8.2 and 8.3 show an astonishing agreement within approx. ±0.05.

Moreover, there is a fundamental difficulty in the interpretation of the results. Granule contours derived from lane maps or iso-intensity cuts through the granulation intensity pattern *do not represent plane cuts* through a temperature field, as would correspond to turbulence experiments performed in the laboratory. Due to vertical motions and radiative cooling processes, iso-thermal layers in the photosphere do not necessarily represent such cuts. Also, different procedures of defining granule boundaries definitely lead to the inclusion of different intensity levels, which again refer to different heights in the atmosphere. Both facts imply that we are dealing with a projection rather than a plane cut. As was explained in Sect. 4, laboratory experiments show that an increase of the thickness of the observed layers tends to yield an over-estimate of the fractal dimension. Therefore, our average value of ≈ 1.9 of the fractal dimension determined with the different methods very probably represents an upper limit, and the question of the 'true' fractal dimension of the granules has to be left with no answer.

9. Conclusion

The determination of the fractal dimension of the solar granulation seemed to be a suitable tool for deciding whether granular structures seen on the solar surface are of turbulent origin or not. From plots similar to our Fig. 4 Roudier and Muller (1986), Darvann and Kusoffsky (1989), and Karpinsky (1990) concluded that granules with a size smaller than about 1.4 arcsec, which seem to have a fractal dimension between 1.14 and 1.25, are of turbulent origin, because the value of d is near that proposed for *iso-bar cuts* through homogeneous, isotropic turbulence. From the literature (Mandelbrot, 1977), however, a value of $2 - \frac{1}{3} = 1.67$ is to be expected for the fractal dimension of the *iso-thermals* determined from plane cuts through such turbulence. Since iso-intensity lines of the granular pattern correspond very closely to *iso-thermal contours*, the finding of d in the range 1.14 to 1.25 in our opinion does not prove the turbulent origin of the smaller granules.

It is important to note that we can indeed reproduce the results of the above authors, even with our somewhat different definition of granules and slightly lower spatial resolution.

A closer inspection of the method in general, however, shows some problems:

- Limited telescopic resolution tends to produce rounder granules in the small and medium granular size range, i.e. diminishes the perimeter for a given area.
- The finite pixel size leads to a preferred fractal dimension around 1.3 for small granules. There are many more small granules than medium size ones. Therefore, the fit is mainly determined by small size granules and even the inclusion of the small number of medium size granules will not change the slope significantly.
- Problems connected to different definitions of the granules also have to be considered.

For small granules the determination of the fractal dimension is dominated by the telescopic resolution on the one and by the pixel size on the other hand.

Our analysis supplies strong arguments for a continuous change of the gradient in the area-perimeter relation with values of d near 1.9 for large granules and questionable values for smaller granules. Moreover, it is reassuring that two additional methods of determining the fractal dimension of granules (i.e. the line and the plane intersection method), applied to both the intensity distribution of the granulation itself and the lane map, all yield very nearly the same result, i.e. a fractal dimension around 1.9.

We believe, that the question of the turbulent origin of the solar granulation can not be answered unambiguously from the determination of the fractal dimension with the present observing material and analyzing techniques .

References

Aharony, A., Feder, J. (1990): *Fractals in Physics* (North-Holland, Amsterdam).

Bradshaw, P., Woods, J.D. (1976): *Turbulence*, (Springer–Verlag, Berlin).

Bray, R.J., Loughhead, R.E., Durrant, C.J. (1984): *The Solar Granulation* (Cambridge Univ. Press, Cambridge).

Carter, P.H., Cawley, R., Licht, A.L., Yorke, J.A., Melnik, M.S. (1986): in *Dimensions and Entropies*, Ed. G. Mayer-Kress (Springer–Verlag), p. 215.

Collados, M., Marco, E., Del Toro, J.C., Vázquez, M. (1986): *Solar Phys.* **105**, 17.

Darvann, T.A., Kusoffsky, U. (1989): in *Solar and Stellar Granulation*, Eds. R. Rutten, G. Severino (Kluwer Acad. Publ., Dordrecht), p. 313.

Deubner, F.L., Gough, D. (1984): *Ann. Rev. Astron. Astrophys.* **22**, 593.

Deubner, F.L. (1988): *Astron. Astrophys.* **204**, 301.

Feder, J. (1989): *Fractals* (Plenum, New York).

Frisch, U., Sulem, P.-L., Nelkin, M. (1978): *J. Fluid Mech.* **87**, 719

Gonzales, R.C., Wintz, P. (1987): in *Digital Image Processing* (Addison-Wesley, New York), p. 331.

Karpinsky, V.N. (1990): in *Solar Photosphere: Structure, Convection, and Magnetic Fields*, Ed. J.O. Stenflo (Kluwer Acad. Publ., Dordrecht) p. 67.

Kolmogorov, A.N. (1941): *C. R. Acad. Sci. URSS* **30**, 299.

Komm, R., Mattig, W., Nesis, A. (1990): *Astron. Astrophys.* **239**, 340

Lovejoy, S. (1982): *Science* **216**, 185.

Mandelbrot, B.B. (1967): *Science* **155**, 636.

Mandelbrot, B.B. (1975): *J. Fluid Mech.* **72**, 401.

Mandelbrot, B.B. (1977): in *Fractals: Form, Chance, and Dimension* (Freeman, San Francisco), p. 278.

Mandelbrot, B.B. (1982): in *The Fractal Geometry of Nature* (Freeman, San Francisco).

Mehltretter, J.P. (1978): *Astron. Astrophys.* **62**, 311.

Nesis, A., Durrant, C.J., Mattig, W. (1988): *Astron. Astrophys.* **201**, 153.

Nesis, A., Mattig, W. (1989): *Astron. Astrophys.* **221**, 130.

Peitgen, H.O., Richter, P.H. (1986): *The Beauty of Fractals* (Springer–Verlag, Berlin).

Roudier, Th., Muller, R. (1986): *Solar Phys.* **107**, 11.

Rüdiger, G. (1989): *Differential Rotation and Stellar Convection* (Akademie–Verlag, Berlin).

Spruit, H.C., Nordlund, Å., Title, A.M. (1990): *Ann. Rev. Astron. Astrophys.* **28**, 263.
Sreenivasan, K.R., Meneveau, C. (1986): *J. Fluid Mech.* **173**, 357.
Stix, M. (1989): *The Sun, An Introduction* (Springer-Verlag, Berlin).
Title, A.M., Tarbell, T.D., Topka, K.P. (1987): *Astrophys. J.* **317**, 892.
Tritton, D.J. (1977): *Physical Fluid Dynamics* (Van Nostrand Reinhold, New York).
Unsöld, A. (1955): in *Physik der Sternatmosphären* (Springer–Verlag, Berlin), p. 219.
Wiehr, E., Kneer, F. (1988): *Astron. Astrophys.* **195**, 310.
Zahn, J.P. (1987): in *Solar and Stellar Physics*, Eds. E.H. Schröter, M. Schüssler (Springer–Verlag, Berlin), p. 55.

Fractals and the Large–Scale Galaxy Distribution

Antonello Provenzale

Istituto di Cosmogeofisica del CNR, Corso Fiume 4, I–10133 Torino, Italy

Abstract: We consider the fractal behavior of the large scale galaxy distribution and we introduce a family of fractal cascading models based on appropriate modifications of the β model of turbulence. These models generate distributions with scale-dependent fractal properties, as observed in the analysis of real galaxy catalogs. In particular, a smooth transition from small-scale self-similarity to large-scale homogeneity is considered. The scale dependence of the fractal properties generates a 'dressing' of the asymptotic value of the fractal dimension to a larger effective value D_{ef}. The physical motivations for the use of a scale-dependent β model in Cosmology are discussed. The multifractal properties of the distributions are considered and an extended version of the β model for the field of density perturbations $\delta\rho/\rho$ is introduced.

I. Introduction

The distribution of luminous matter in the Universe is far from being homogeneous, at least for scales up to some tents of Megaparsecs. Individual galaxies are tied up into groups and clusters, the latter containing up to hundreds of galaxies. In turn, clusters are often grouped into superclusters, which are among the larger individual structures that have been detected in the Universe. On the other side of the spectrum, there are large underdense regions, called voids, which display a very low density of luminous objects. The observed distribution of galaxies is apparently very irregular and intermittent, and it possesses clear hierarchical properties, as evidenced for example by the peculiar behavior of the n-point galaxy-galaxy correlation functions.

A quantitative approach to the study of hierarchical distributions is based on the use of fractals, mathematical objects which naturally describe the irregularity and the scaling behavior found in many physical systems. For these reasons, several attempts have been made in order to characterize and modelling the large-scale distribution of matter in terms of fractal distributions. The analysis of the fractal properties of the available galaxy and cluster catalogs has confirmed that the galaxy distribution may be represented as a fractal dust, as already indicated by the power-law form of the two point correlation function. Here we briefly review some of the concepts underlying the application of fractals in Cosmology and we discuss a class of fractal cascading models which describe the large-scale galaxy distribution.

II. The Fractal Dimension of the Galaxy Distribution

A correct definition of a fractal set is 'a mathematical object whose fractal (Hausdorff) dimension D is strictly larger than its topological dimension D_T.' The rigorous definition of Hausdorff dimension may be found for example in Mandelbrot (1982), here we use the common definition based on the box-counting algorithm. The box-counting dimension D_0 (in the following called the 'fractal dimension' tout court) of a point distribution is given by

$$D_0 = \lim_{r \to 0} \frac{\log N(r)}{\log 1/r} \tag{1}$$

where $N(r)$ is the number of cubes with side r which are needed to cover the distribution.

An important property of fractal distributions is their scaling behavior. In this context, the term scaling indicates that the structures present at a certain scale (e.g. the density fluctuations) are related to the corresponding structures at the other scales by a general recursive relation. A fractal set may be deterministic, i.e. it may be generated by a deterministic algorithm (like a Koch curve or a Cantor set), or may be a stochastic fractal associated with the output of a stochastic process. In the latter case, the scaling behavior refers to the statistical properties of the fractal set. In general, the fractals which are commonly used in Physics and Cosmology are stochastic fractals. An important class of fractals are self-similar mono-fractals, which are characterized by a single fractal dimension. These fractal sets are characterized by the fact that every part of the set is an exact replica of the whole set (in a statistical sense). In this case, all moments of the probability distribution scale with the same scaling exponent. More complex fractal sets are represented by the so-called multifractals, see below. For a multifractal, the different moments of the probability distribution scale with a different scaling exponent. This implies that peaks of different strength in the distribution (e.g. peaks of different density) possess *different* scaling properties (i.e. they have a different fractal dimension).

The use of formula (1) is not the only method for computing an approximation to the Hausdorff dimension. In particular, another useful algorithm is based on the calculation of the correlation integral $C(r)$ introduced by Grassberger and Procaccia (1983) in the study of strange attractors. The correlation integral $C(r)$ is given by

$$C(r) = \lim_{N \to \infty} \frac{1}{N^2} \sum_{i,j=1}^{N} \Theta\left(r - |\boldsymbol{X}_i - \boldsymbol{X}_j|\right) \tag{2}$$

where Θ is the Heaveside step function and N is the total number of points in the distribution; $C(r)$ measures the probability that two randomly selected points with positions $\boldsymbol{X}_i$ and $\boldsymbol{X}_j$ be closer than the distance r. This quantity measures also the probability of finding another point inside a sphere with radius r, centered on a generic point of the distribution. $C(r)$ thus represents a two-point probability. For a fractal distribution it is possible to show that

$$\lim_{r\to 0} C(r) = r^{D_2} \tag{3}$$

where D_2 is the correlation dimension of the point set; D_2 is again an approximation to the Hausdorff dimension D of the set. Apart from statistical fluctuations, for simple fractals one has that $D_2 = D_0$, while for multifractals $D_2 < D_0$. In general, D_2 provides a lower limit to the Hausdorff dimension. For a point distribution ('dust') the topological dimension is D_T=0. For a fractal dust the fractal dimension D must larger than 0, homogeneity corresponds to D=3. By plotting $\log C(r)$ versus $\log r$ one immediately realizes whether the correlation integral possesses a power-law behavior. The correlation dimension is then obtained as the slope of the least-squares-fit line to $\log C(r)$ versus $\log r$ or as the average value of the local logarithmic slope of the correlation integral.

The correlation integral $C(r)$ is related to the two-point correlation function, a classic quantity in Cosmology. The two point correlation function $\xi(r)$ is defined by the expression (Peebles 1980)

$$N\delta P = n\,(1+\xi(r))\,\delta V \tag{4}$$

where δP is the average probability of finding a galaxy in the small volume δV placed at a distance r from a generic galaxy of the sample; n is the average number density of galaxies and N is the total number of galaxies in the sample. Clearly, the correlation function (4) represents a two-point probability; it measures the probability *in excess* with respect to a random homogeneous distribution of finding a galaxy in the volume δV. For a random, homogeneous distribution of points one has in fact that $\xi(r)$=0. If the positions of the galaxies are correlated, then $\xi(r) > 0$, while if the galaxy positions are anticorrelated it is $\xi(r) < 0$. It is easy to see that the correlation function $\xi(r)$ may be obtained from the derivative of the correlation integral $C(r)$, i.e.

$$1+\xi(r) = \frac{N}{4\pi n r^2}\frac{dC(r)}{dr} \tag{5}$$

From a practical point of view, the two-point correlation function is often computed as the ratio of the number of galaxies $\Delta N(r)$, placed in the volume ΔV at distance r from a reference galaxy, over the number of points $\Delta N_R(r)$ which would be find in the volume ΔV for a corresponding random homogeneous distribution, i.e.

$$1+\xi(r) = \frac{\Delta N(r)}{\Delta N_R(r)} \tag{6}$$

From expression (5), it is also clear that

$$1 + \xi(r) = \frac{dC(r)/dr}{dC_R(r)/dr} \tag{7}$$

where $C_R(r)$ is the correlation integral for the random distribution. Operationally, given a point distribution and a set of uniform distributions, one first computes the quantities ΔN and ΔN_R, the latter as given by the average over the ensemble of random distributions. From the above quantities, the correlation function is then obtained by using expression (6) and the correlation integral $C(r)$ is obtained by integrating $\Delta N(r)$ from zero to r. When studying real galaxy catalogs, several problems generated by the limited number of galaxies in the catalog may be encountered. For these reasons, several methods to compute the two-point correlation function from available galaxy catalogs have been proposed (see e.g. Davis and Peebles 1983; Rivolo 1986; Blanchard and Alimi 1988). In particular, attention must be paid to the use of volume-limited or magnitude-limited samples, as well as to a careful evaluation of statistical uncertainties and of boundary effects. Clearly, these problems are much less relevant in the analysis of syntethic galaxy catalogs like those considered here.

In the case of a fractal distribution of points with correlation dimension D_2, one has that $C(r) \approx r^{D_2}$ and consequently, from formula (5), that $1 + \xi(r) \approx r^{D_2-3}$. Obviously, when $\xi(r) \gg 1$ one has that $\xi(r) \approx r^{D_2-3}$. It is thus clear that a power-law form of the correlation function is an indication of scale-invariant clustering and of fractal behavior. An important observational result is that the real galaxy distribution has a correlation function with a power-law dependence, i.e. $\xi(r) \approx (r/r_0)^{-\gamma}$, with γ approximately 1.7 or 1.8 and r_0 of the order of 5 Mpc, for distances up to about ten Mpc (see e.g. Peebles 1980). This result implies a correlation dimension D_2 of about 1.2 or 1.3 for the galaxy distribution at these scales. At larger scales, the analysis of cluster catalogs has revealed an analogous scaling behavior with apparently the same correlation dimension as the galaxy distribution (see e.g. Bachall 1988). These results would thus suggest a self-similar fractal behavior of the distribution of luminous matter in the Universe. It is also important to recall that the relevant quantity for the study of the fractal behavior is the function $1 + \xi(r)$, which is immediately related to the correlation integral and thus to the value of the correlation dimension D_2. On the contrary, the usual quantity adopted in the study of the galaxy distribution is $\xi(r)$. This difference is of no importance when the correlation function is much larger than one, but it may become relevant when studying scaling regimes at large spatial separations, where $\xi(r)$ is order one or less. See below for a discussion of this issue.

Beyond the above results, however, several issues have still to be properly addressed. The nature of the hierarchical behavior of the galaxy distribution is still widely debated. For example, Mandelbrot (1982), Pietronero (1987) and Coleman et al. (1988) have suggested that the distribution of galaxies may be described as a simple self-similar fractal (an homogeneous fractal dust), extending over all length scales. However, if the large scale matter distribution is a pure self-similar fractal, then a meaningful average density of the Universe cannot be defined, since for a pure fractal the average density is vanishing in the limit of an infinite sample volume. Practically, by considering larger and larger sample volumes one would find an ever decreasing average density of the Universe (Pietronero 1987; Calzetti et al. 1988). For this reason, the idea of a mono-fractal distribution of galaxies, with no upper homogeneity cutoff, is not easy to accept. In fact, this is apparently in contrast with both the uniformity of the cosmic microwave background and with standard big bang theories. In addition, Peebles (1988; 1990) has shown that the present data do not support a simple self-similar distribution of visible matter, and the most recent analyses of the available galaxy catalogs (see e.g. Martinez 1990) seem to indicate the existence of a transition from small-scale fractal behavior to a large-scale homogeneous distribution. In this case, the galaxy distribution would possess scale-dependent fractal properties, approaching homogeneity at the largest scales. An average matter density could thus be meaningfully defined for scales larger than the homogeneity scale. In addition to the scale-dependent fractal behavior, Jones et al. (1988) and Martinez and Jones (1990), in an analysis of the CfA galaxy catalog, have found apparent evidence of a multifractal behavior of the galaxy distribution.

In order to better understand the fractal nature of the galaxy distribution, several phenomenological models based on the use of fractal algorithms have been proposed, see e.g. Soneira and Peebles 1978; Schulman and Seiden 1986; Vicsek and Szalay 1987; Liu and Deng 1988). These models, however, are in general based on the notion of self-similarity and of unlimited scaling behavior. Simple fractal models of the galaxy distribution with an upper homogeneity cutoff have been proposed as well (Ruffini et al. 1988; Calzetti et al. 1988). In these models, the matter is supposed to be distributed in large-scale cells which have a random homogeneous distribution. Inside the cells, a self-similar cascading process is activated. The matter distribution is thus homogeneous at scales larger than the size l_0 of the cells and it is a pure fractal for scales smaller than l_0. A simplified assumption of these models is that they consider only two regimes (self-similarity and homogeneity), with a sharp transition from one regime to the other.

The required homogeneity at large scales, together with the need of extending the notion of self-similarity, suggests on the other hand that it may be useful to consider models whose scaling and fractal properties vary with the spatial scale. For example, Lukash and Novikov (1988), stimulated by the results of Jones et al. (1988), have suggested that the fractal dimension of the galaxy distribution may possess a physically motivated scale-dependence. In this view, the fractal dimension grows from D=1 at small scales (a value which is consistent with the result of nonlinear gravitational clustering and with the presence of filamentary structures), to D=2 at intermediate scales, where the presence of pancakes dominates the matter distribution (see e.g. Shandarin and Zeldovich 1989), to D=3 at the largest scales, where the distribution becomes homogeneous. In this regard, we recall that the fractal (Hausdorff) dimension is rigorously defined only in the limit of small scales, but nevertheless different 'characteristic' fractal dimensions (or better 'characteristic fractal exponents') may be properly associated with different scale ranges if a local scaling behavior is observed, in the spirit of 'intermediate asymptotics' (see e.g. Barenblatt 1980). Along these lines, we have introduced a family of simple fractal cascading processes which generate an homogeneous distribution at the largest scales and which provide a smooth transition from fractality to homogeneity (Castagnoli and Provenzale 1990, 1991; Provenzale 1990). These models are based on appropriate modifications of the β model of turbulence, which is briefly reviewed in the next section.

III. The Self–Similar β Model

The basic version of the β model has been introduced some years ago in the study of turbulence (Frisch et al. 1978). The simplest realization of the β model is a three-dimensional distribution of points generated by a sequence of breaking iterations starting from a parent object (e.g. a cube) with linear size l_0 (for simplicity, l_0=1) into M smaller objects with linear size $l_1=l_0/M^{1/3}$. For simplicity, from now on M=8 and l_1/l_0=1/2. Of the M 'son' objects, only m objects remain 'active' and are able to break again at the second iteration. The number of active objects is then taken to be a random variable whose mean value is fixed by the 'survival probability' p of each cube, i.e. $< m > = pM$. This must be understood in the sense that each object has probability p of breaking again (of 'surviving') and probability $1-p$ of becoming 'quiescent'. In the simplest version of the model, the probability p is equal for all objects and for all iterations. By repeating several times this procedure, a fractal distribution of points is obtained. The fractal dimension of the distribution may then be found from definition (1), that is

$$D_0 = \lim_{k\to\infty} \frac{\log\ < m >^k}{\log\ 2^k} = \frac{\log\ < m >}{\log\ 2} = \frac{\log\ pM}{\log\ 2} \tag{8}$$

since $r = l_k = 1/2^k$ and $N(r) = m^k$, where the index k denotes the k-th iteration in the breaking cascade. The value of the fractal dimension is completely determined by the value of the survival probability p.

The distributions generated by the basic version of the β model are self-similar, i.e. no upper homogeneity cutoff is present. For this kind of distributions, the fractal behavior is scale-independent and it extends over all length scales. The size of the voids (i.e. of the regions devoid of active objects) increases with the sample size, and the matter density vanishes in the limit of a sample of infinite volume. While the basic version of the β model may have some interest for describing the galaxy distribution at small scales, it is presumably inadequate for the purpose of modelling the behavior of the matter distribution at all scales. In Sect. V we consider a simple modification of the β-model which may be used to describe both the transition to large-scale homogeneity and the scale-dependence of the fractal properties of the galaxy distribution. In the next section we discuss the physical motivations underlying the introduction of a (modified) cascading β model in Cosmology.

IV. Physical Motivations of a 'Cosmological' β Model

The β model discussed in the previous section has been developed as a phenomenological description of the energy cascade in fully developed turbulent flows. In the study of turbulence the 'active objects' must be interpreted as the active regions of dissipation, or, alternatively, as the dynamically active turbulent 'eddies' (vortices) in the energy cascade. The survival probability must be interpreted in terms of the probability of an eddy of being active. The fact that not all eddies 'survive' means that the region of space occupied by active eddies is not space-filling. Note that the distribution generated by a β model may be used to describe both direct energy cascades (as in three dimensional turbulence) or inverse energy cascades (as in two dimensional turbulence). In a cosmological context, however, the interpretation of the β model in terms of turbulent cascades is not completely clear. In the applications to Cosmology there are however at least two alternative scenarios in which the introduction of the cascading β model turns out to be quite natural. In each context, the meaning of the terms 'active objects' and 'survival probability' must be defined.

A first scenario is in terms of hierarchical gravitational clustering. In a cold dark matter (CDM) Universe, the spectrum of the density fluctuations after recombination has most power in the small scale range (see e.g. White 1987). In this case, small scale structures form first; the larger structures are formed by subsequent gravitational clustering of smaller clumps of matter. This is known as the 'bottom-up' scenario; in the terminology of the β model this corresponds to an inverse energy cascade. A hierarchical nature of the distribution of the clustered matter is then theoretically suggested by the lack of characteristic length scales in the equations for gravitational clustering. This in turn suggests the existence of a self-similar galaxy distribution, at least on the scales affected by 'fully-developed' (nonlinear) gravitational clustering.

An obvious question in this context concerns the value of the fractal dimension of the gravitationally clustered distribution. Clearly, it is possible to use the classic β model in order to generate a distribution with the observed value $D \approx 1.2$. However, there is no theoretical reason to select such a value for the fractal dimension. On the other hand, a value $D=1$ for the fractal dimension of the clustered matter has been theoretically obtained in different contexts. This result has been obtained either by considering the finiteness of the peculiar velocities of virialized astronomical objects (Fournier d'Albe, 1907) or by dimensional reasons based on the definition of the Jeans mass (Hoyle 1953), see Mandelbrot (1982) for a review. The argument based on the finiteness of peculiar velocities proceeds as follows. In a *virialized* object with mass $M(R)$ and typical radius R, the potential energy is balanced by twice the kinetic energy, i.e., apart from numerical constants, $M(R)V^2 \approx GM(r)^2/R$ where G is the gravitational constant and V is a typical r.m.s. velocity of the object. Thus, if we want V^2 to be finite for larger and larger values of R, it must be $M(R) \approx R$, i.e. the mass distribution must have a fractal dimension $D=1$ (recall Eq. (3)). Clearly, this argument is applicable only for objects which are virialized (or at least gravitationally bound), i.e. objects which have undergone a 'fully–developed' gravitational clustering. We also recall that, more recently, a value D=1 for the matter distribution has been obtained in the context of cosmic string scenarios, see e.g Turok (1986), and Schramm (1988), for reviews. The dimensional arguments, however, lead to a stricly self-similar distribution of matter, at least for virialized objects. Given the large scale homogeneity, we know on the other hand that self-similarity must be broken at a certain scale, related for example to the size of the objects which have not been yet virialized.

Considering the dynamical evolution of the density field, we note that in the CDM scenario the required breaking of the self-similar clustered distribution has a natural physical origin. In a CDM Universe, the gravitational clustering begins in fact only after the recombination epoch. On the other hand, the process of gravitational clustering takes a certain time to increase the density fluctuations, to reach the nonlinear stage and finally to generate virialized objects. Since in a CDM Universe the clustering begins from the small scales, it is clear that a self-similar behavior at all scales is reached (if ever) only after a sufficiently long time, when also the largest scales have undergone gravitational clustering and virialization. As a consequence, at any finite time a smooth transition from small-scale fractality (produced by the clustering) to large scale homogeneity (the initial state) is in general observed. This indicates that the field of density fluctuations $\delta\rho/\rho$ has different scaling properties at different scales, becoming more irregular and intermittent at the smallest scales. The field $\delta\rho/\rho$ is thus linked with a fractal dimension D=1 at small scales and with a dimension D=3 at large scales, where the fluctuations have not yet grown enough and the field is still essentially homogeneous. In this sense, the appearance of a fractal dimension D=1 for the matter distribution would be associated with the objects which have fully entered the nonlinear regime.

A natural representation of the above process may be obtained by using a cascading β model with a scale-dependent fractal behavior. Since in the β model the value of the fractal dimension is fixed by the survival probability, a possible way to obtain scale-dependent fractality is to consider a survival probability $p(l_k)$ which depends upon the scale l_k of the objects at the k-th iteration. In the CDM interpretation, even though the β model is practically built as a cascade from large to small scales, the physical construction of the galaxy distribution proceeds from small to large scales. This corresponds to an inverse cascade in the terminology of turbulence. The simplest distributions are thus obtained by considering a survival probability p=1 (corresponding to homogeneity and to a dimension D=3) at large scales (larger than, say, an 'homogeneity scale' l_0) together with a well-defined probability p_0=1/4 at the scales of virialized objects. This value of p_0 fixes through formula (8) the asymptotic value D_{as}=1 for the fractal dimension of the matter distribution at small scales. A smooth transition between these two scaling regimes is then generated. Presumably, the details of the transition region are less important. See however Sect. VI for a discussion on the possible presence of another scaling regime at intermediate scales. We recall that the type of model introduced by Calzetti et al. (1988) may be interpreted in terms of a β model with a sharp transition from large-scale homogeneity to small-scale self-similarity, without any 'transition region'. This region is important, however, since it may generate a 'dressing' of the small-scale fractal dimension D_{as} to an effective value $D_{ef} > D_{as}$, see the next section.

In the context of the CDM scenario, the term 'active objects' used in the β model must be interpreted in terms of gravitational clumps, i.e. in terms of positive density fluctuations which have undergone or are undergoing gravitational clustering. The 'survival probability' $p(l_k)$ determines the average number of 'sons' present inside a larger object; physically it is related to the amount of space occupied by substructures inside a larger structure. A probability $p = 1$ implies that the substructures fill completely the large structure, i.e. that the substructures have not reached the state of 'individual' entities and that the density field does not have strong peaks. A probability smaller than one implies that the substructures do not fill completely the larger structure, i.e. that gravitational clustering is active and that the density field possesses both strong peaks and underdense regions. The probability descreases as the gravitational clustering becomes more and more efficient, i.e. as the fluctuations becomes more and more nonlinear. We note that in this approach, the form of the survival probability $p(l_k)$ must be time-dependent, being associated with a continuous growth of the clustered (nonlinear) region of scales as time goes on.

A second cosmological scenario which may naturally lead to the cascading β model is the fragmentation of large scale structures (e.g. pancakes) in a hot dark matter (HDM) Universe (see e.g. Shandarin et al. 1983). In this context, large-scale structures form first, followed by a complicated fragmentation process which is dominated by both gravitational instability and by hydrodynamical processes. In this case, the β model may be used for parametrizing the fragmentation of the large scale structures, and again a survival probability depending upon the scale is required. For example, one of the motivations for such a dependence is found in a scale-dependent balance between gravitational instability and hydrodynamical effects. This cosmological scenario is discussed in more detail in Castagnoli and Provenzale (1991). Here we recall that in this case, the cascade must be interpreted as a direct cascade (from large to small scales), and that the subsequent iterations in the fragmentation process take place at different times, contrarily to what happens in the CDM scenario.

As a conclusion to this section, we stress the fact that there are physical motivations which suggest the use of a cascading β model as a descriptor of the large scale distribution of luminous matter. However, the model still retains its fundamental phenomenological nature, and further work is required in order to obtain a fully consistent picture. Along these lines, we are now considering a comparison between the distributions generated by a cascading β model and by N-body numerical simulations. As a first step, we have studied the evolution of the fractal properties of the distributions generated by N-body simulations (Valdarnini et al. 1991).

Another line of thought, however, is to consider the β model with scale-dependent fractal behavior as a 'fractal machine' which is capable of producing point distributions with controllable fractal properties. Such a tool may be very useful for testing the results of data analysis methods that are sometimes difficult

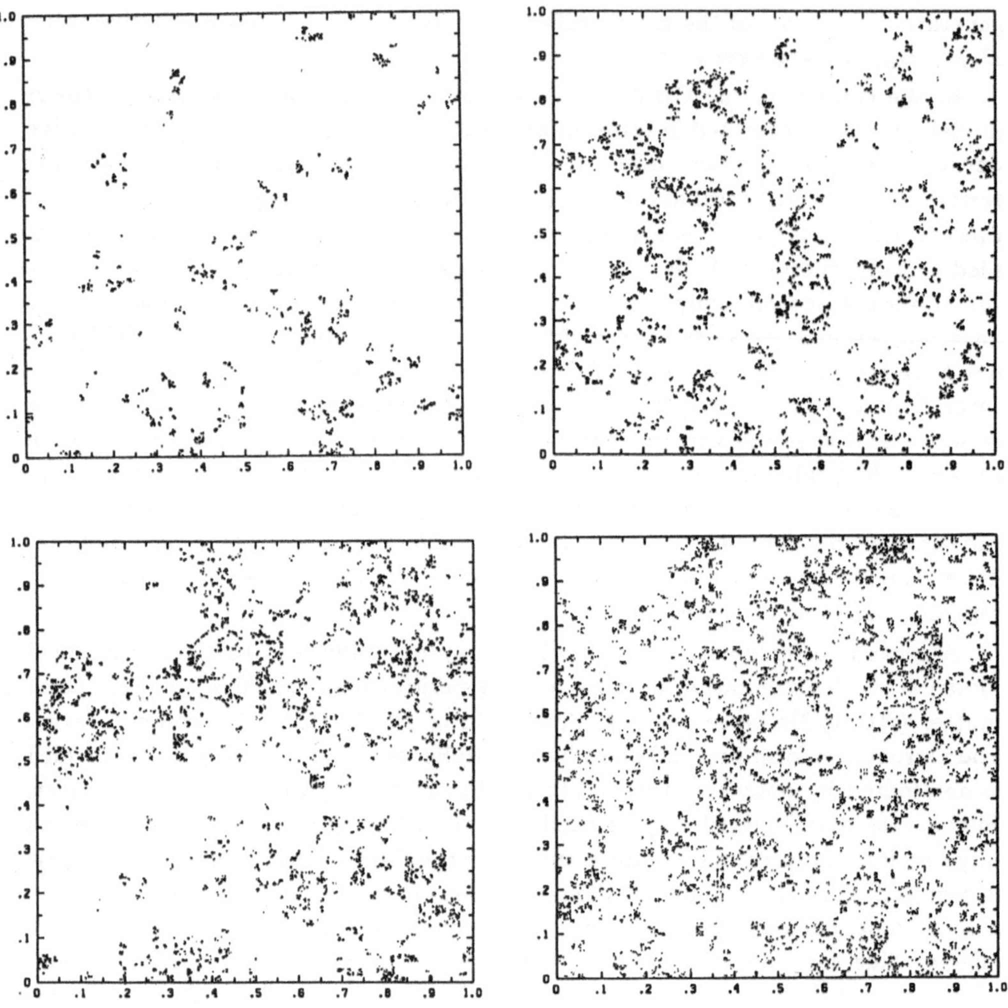

Fig. 1. Slices of four three-dimensional distributions generated by the fractal cascading model discussed in Sect. V. The four panels correspond respectively to the choices $L=1/4$, $L=1/8$, $L=1/10$ and $L=1/12$. The slices have a thickness which is one tenth of the size of the basic cell.

to interpret in the study of real galaxy catalogs (owing to the limited statistics, for example). In this case, the β model (and subsequent modifications) may provide a 'reference point' for the analysis of the fractal behavior of the galaxy distribution. Analogously, these models may be used as an efficient way of compacting the relevant information contained in the hierarchical behavior of the galaxy distribution, or analogously as the 'minimal' models capable of reproducing the observed fractal behavior. Along these lines, we are now considering the inverse problem of deriving the optimal form of the survival probability $p(l_k)$ from the analysis of the observed large scale galaxy distribution.

V. The Scale–Dependent β Model

As mentioned above, the simplest way of generating a point distribution with an upper homogeneity cutoff is to consider an abrupt jump from self-similarity to homogeneity. In the framework of the β model this would correspond to a jump of the survival probability from a value $p = 1$ for scales larger than the homogeneity scale l_0 to a value $p < 1$ for scales smaller than l_0. A simple way of generating both a smooth transition from fractality to homogeneity and a scale-dependent fractal behavior is to allow the survival probability p to be a smooth function of the size l_k of the breaking object, i.e. $p = p(l_k)$.

Clearly, the form of the function $p(l_k)$ must be given by a detailed physical mechanism, associated with both the oucome of gravitational clustering and with the spectrum of density fluctuations after recombination. See Sect. VII for a discussion of this issue. Following for now a purely heuristic approach, we consider only the two basic physical requirements that D=3 (i.e. p=1) at large scales and that D_{as}=1 (i.e. p=1/4) at small scales, and we consider as an example the form of $p(l_k)$ given by

$$p(l_k) = 1 \qquad for \qquad l_k \geq l_0$$

$$p(l_k) = p_0 + (1 - p_0)\ exp\left(L(\frac{1}{l_k} - \frac{1}{l_0}\right) \qquad for \quad l_k < l_0 \tag{9}$$

where p_0=1/4 and l_0 is the size of the largest objects considered (well inside the homogeneity region), hereinafter l_0=1. The parameter L fixes the smoothness of the transition from homogeneity to fractality. Expression (9) has been chosen mainly because it is one of the simplest examples of a smooth transition from p=1 at large scales to a well-defined, constant value p_0 at small scales. Other choices of similar smooth functions lead to analogous results. The parameter L fixes also the width of the transition region; this parameter must be time-dependent in the interpretation of the β model in the CDM scenario. For L going to zero one obtains a homogeneous distribution; a pure fractal set with dimension D=1 is found in the limit for L going to infinity.

Fig. 1 reports the 'slices' of four three-dimensional distributions obtained from the β model with $p(l_k)$ given by formula (9) with L=1/4, L=1/8, L=1/10 and L=1/12 respectively. Each three-dimensional distribution contains approximately 100,000 galaxies; the thickness of each 'slice' is one tenth of the basic cube. The distributions corresponding to L=1/8 and L=1/10 closely resemble the observed galaxy distribution, with clusters, filaments and voids.

Fig. 2 reports the correlation integral $C(r)$ for the three-dimensional distributions reported in Fig. 1 and for a uniform, random distribution. A well-defined fractal behavior is present at intermediate scales; a transition from fractality to

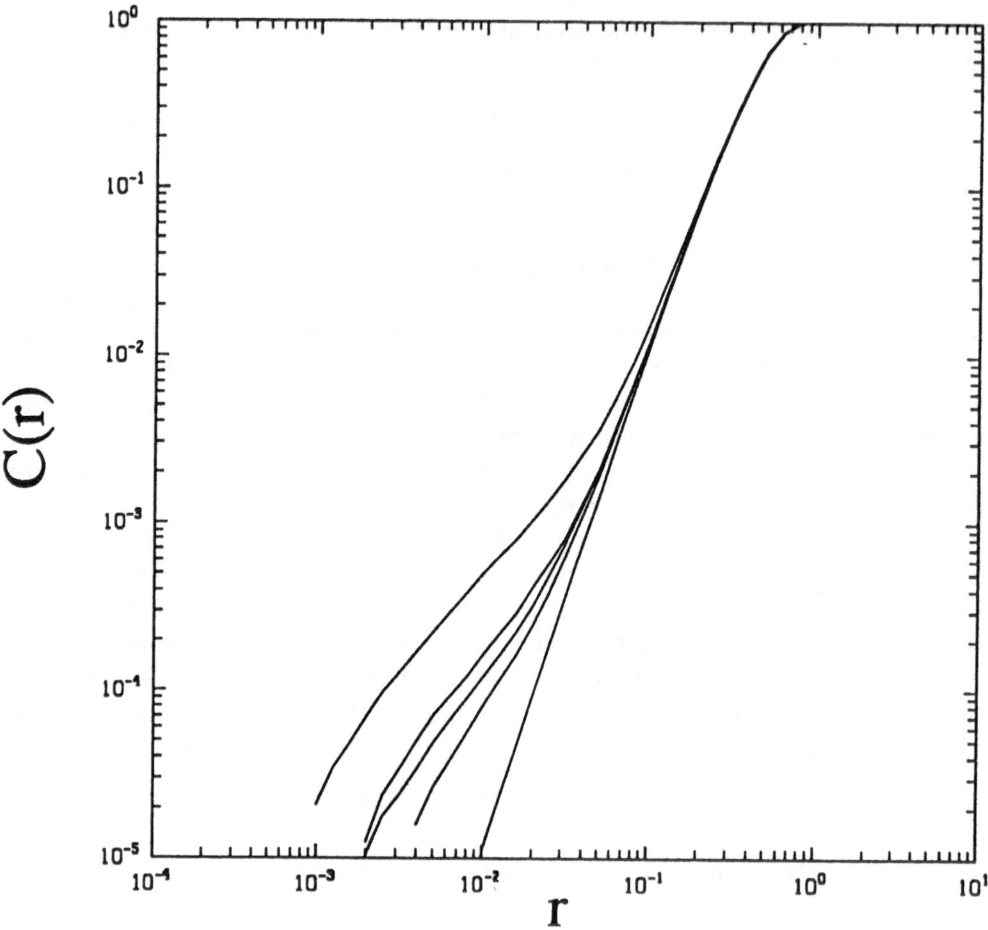

Fig. 2. Correlation integrals for the four galaxy distributions shown in Fig. 1 and for a random homogeneous distribution. The values of L corresponding to each curve are decreasing from left to right. The last curve on the right is for the homogeneous distribution.

homogeneity is seen at scales comparable with the value of L. The best-fit logarithmic slope of the correlation integral at scales smaller than L furnishes a correlation dimension D_2=1.10 for L=1/4, D_2=1.23 for L=1/8, D_2=1.32 for L=1/10 and D_2=1.58 for L=1/12. At larger scales, the slope of $\log C(r)$ versus $\log r$ tends to the value $D_{hom} \approx 3$, typical of uniform random distributions.

It is important to note that the value of the computed correlation dimension D_{ef} is always larger than the asymptotic value (D_{as}=1), the difference becoming larger as L decreases. In fact, the detailed dependence of the survival probability upon the spatial scale affects the value of the 'effective' dimension D_{ef} which is practically measured in a finite sample. This 'dressing' mechanism may be responsible for the difference between the computed value of the correlation dimension

$D_2 \approx 1.2$ and the theoretical value $D=1$ discussed in the previous section. In fact, values of L about $1/10\ l_0$, corresponding to distributions which are visually similar to the real galaxy catalogs, provide precisely a dressing of the dimension from the value $D=1$ to a value $D_{ef} \approx 1.2$.

Fig. 3 reports the two-points correlation function for the four galaxy distributions shown in Fig. 1. A power-law dependence is observed at intermediate scales, while $\xi(r) \approx 0$ at scales larger than the homogeneity scale. In order to provide a physical meaning to the quantities involved in the β model, we note that the physical value of the size l_0 may be fixed at approximately 200 Mpc. The correlation length of the galaxy distribution (i.e. the value of r at which $\xi(r)=1$) is then found at scales between 4 and 6 Mpc, in good qualitative agreement with the observations. The homogeneity scale is then found between about 30 and 100 Mpc, depending upon the value of L. For these distributions the dimension of the voids does not increase without bound, and a meaningful average density can be defined at sufficiently large scales.

The distributions generated by the β model considered above possess scale-dependent fractal properties, as apparently observed in the analysis of the real galaxy distribution. An additional issue concerns the multifractal behavior of the galaxy distribution. Multifractal sets are characterized by the fact that the different moments of the distribution scale with a different fractal behavior. This leads to the introduction of a spectrum of generalized fractal dimensions, associated with the moments at different orders. For simplicity, here we consider the definition of the generalized dimensions based on a simple extension of the Grassberger and Procaccia algorithm. Further details on multifractal sets may be found for example in Paladin and Vulpiani (1987) and in Martinez et al. (1990). The n-th order correlation integral is defined by

$$C^{(q)}(r) = \frac{1}{M} \sum_{i=1}^{M} [C_i(r)]^{q-1} \tag{10}$$

where M is the number of sample galaxies which have been randomly extracted from the distribution; obviously M is less or equal to the number N of galaxies. The correlation integral for the i-th sample galaxy is given by

$$C_i(r) = \lim_{N\to\infty} \frac{1}{N} \sum_{j=1}^{N} \Theta\left(r - |\boldsymbol{X}_i - \boldsymbol{X}_j|\right) \tag{11}$$

For a (multi)fractal distribution one has

$$\lim_{r\to 0} C^{(q)}(r) = r^{(q-1)D_q} \tag{12}$$

where D_q is the q-th generalized dimension. The correlation dimension is found for $q=2$ and the box-counting dimension corresponds to $q=0$. For monofractals, all the generalized dimensions are equal. For multifractals, one has in general that $D_q > D_{q'}$ if $q' > q$.

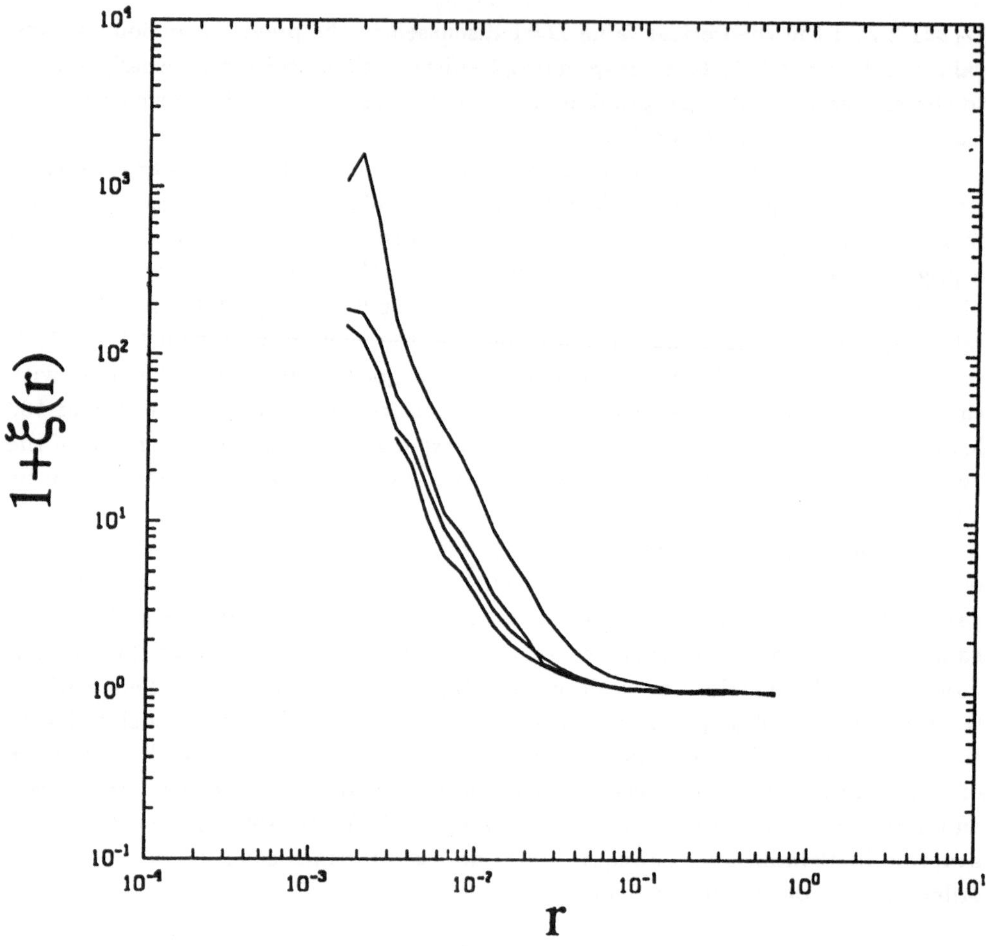

Fig. 3. Two-point correlation functions for the galaxy distributions shown in Fig. 1. The values of L for each curve are decreasing from top to bottom.

The evaluation of the spectrum of generalized dimensions for the distributions generated by the cascading β model reveals an essentially monofractal behavior of these point distributions. This is easily understood since a multifractal behavior can be obtained only by a multiplicative random process, while the β model is associated with an additive process. These results indicate that scale-dependent fractality and multifractality are not necessarily associated with each other. For the β model, in fact, one observes a scale-dependent mono-fractality. In the analysis of the CfA catalog, on the other hand, some evidence has been found that the galaxy distribution may possess a multifractal nature (Jones et al. 1988). However, we note that the spectrum of generalized dimensions has been evaluated from a very limited number of objects (between three and four hundred galaxies), while a much larger number of points is required in order to provide a stable and correct estimate of the spectrum. Another evidence of multifractality comes from the difference between the dimension estimate given by a minimal spanning tree (MST) algorithm and the correlation dimension (Martinez and Jones 1990).

The MST dimension D_{MST} is supposed to provide an estimate of the Hausdorff dimension, which is close to D_0, while the correlation dimension corresponds to D_2. A value $D_{MST} \approx 2$ has been obtained, to be compared with the known value $D_2 \approx 1.2$. This result has been interpreted as an evidence of multifractality. However, we note that the difference between these two dimensions is quite large, implying a very wild multifractal behavior. In addition, it is still not clear how the different methods are sensitive to a possible scale-dependence of the fractal behavior of the galaxy distribution, especially in the case of very limited samples. The MST method, for example, could be more sensitive to the large scales, where a larger fractal dimension (a 'more homogeneous' distribution) may be present. In addition, the fact that individual galaxies and clusters (presumably associated with peaks of different strength in the density field) scale similarly is also suggestive of monofractality. In fact, a multifractal distributions is characterized by the fact that peaks of different strength scale with a different fractal dimension. We also note that the outcome of gravitational clustering is supposed to provide a self-similar (i.e. monofractal) distribution of virialized objects. If this is the case, then multifractality could be found only in the transition region at large scales. In Sect. VII we introduce a modification of the β model which is capable of furnishing small-scale monofractality and large scale multifractality, together with a smooth transition to homogeneity. In any event, further theoretical work and data analysis efforts are clearly required on these intriguing issues.

VI. A 'Staircase' β Model

In the scale-dependent β model discussed in the previous section, we have supposed that only two scaling regimes were present, namely small-scale self-similarity, generated by gravitational clustering and associated with a dimension D_{as}=1, and large scale homogeneity, associated with a dimension D_{hom}=3. The transition region has been supposed to be non-scaling. However, recent analyses of the Perseus-Pisces region redshift data (Guzzo et al. 1990), have suggested the existence of two scaling regions in the large scale distribution of galaxies, associated with a dimension D=1 at small scales (generated by the gravitational clustering), and with a dimension D=2 at larger scales. A transition to homogeneity is then supposed to be present at still larger scales. Note that this result is quite consistent with the suggestion of Lukash and Novikov (1988) on the existence of a scale-dependent fractality in the galaxy distribution, with a fractal dimension varying from=1, to D=2 and finally to D=3.

The results concerning the presence of two scaling regimes in the galaxy distribution have been obtained by Guzzo et al. (1990) by considering the behavior of the two-point correlation function. An important observation made in that work is that the two scaling regimes are clearly detectable only when the function $1+\xi(r)$ is used. When considering the function $\xi(r)$, the second scaling regime at large spatial separations is lost, the only remainder of this second region being a 'feature' which perturbs the power-law behavior of the correlation function at large scales. On the other hand, the correlation function $\xi(r)$ is not related to the fractal and scaling behavior of the galaxy distribution; for the study of fractal behavior the appropriate quantity is thus $1 + \xi(r)$, which is related to the correlation integral $C(r)$.

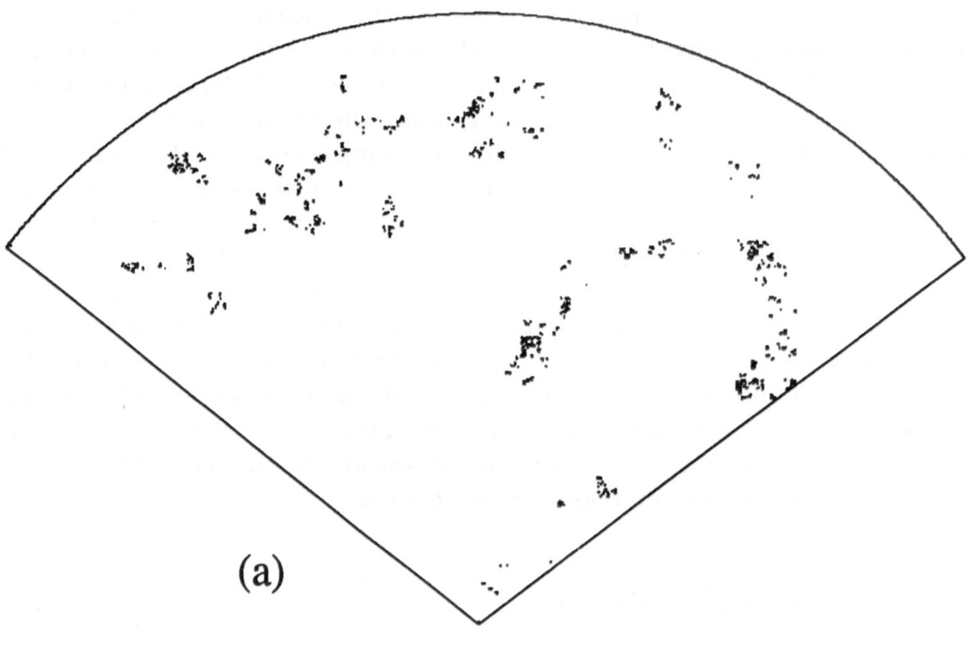

Fig. 4a. A slice of a three-dimensional galaxy distribution generated by the 'staircase' β model discussed in Sect. VI.

Starting from the above results, we have used the cascading β model to generate a galaxy distribution with the same characteristics as apparently observed in the data (see e.g. Provenzale 1990). This approach is discussed in detail in a forthcoming paper (Guzzo and Provenzale, in preparation), here only a brief introduction to the subject is given. Fig. 4a reports a slice of the distribution of galaxies generated by the cascading β model with survival probability p=1 at large scales, p=1/2 at intermediate scales (corresponding to D=2), and p=1/4 at small scales. The slice is obtained by selecting a spherical sector centered on a galaxy of the sample; a physical depth of the sample may be fixed at about 100 Mpc. Fig. 4b reports the function $1 + \xi(r)$ (upper curve) and $\xi(r)$ (lower curve) for the three-dimensional distribution. Two well-defined scaling regimes are evident in the behavior of $1 + \xi(r)$; the transition between the two regimes is found between approximately two and five Mpc. The slope of the correlation function is $\gamma \approx 2$ at small scales and $\gamma \approx 1$ in the second scaling regime. At large scales, the function $1+\xi(r)$ tends to one, revealing the presence of homogeneity. On the other hand, in the function $\xi(r)$ the second scaling regime is almost completely lost; only a small perturbation to the power-law behavior is left. In this context, the cascading β model may consequently be used as a quantitative tool for studying the properties of real galaxy catalogs and for testing data analysis methods on synthetic catalogs with controllable fractal behavior. In particular, the above results confirm that in the study of the fractal and scaling nature of the galaxy distribution at large scales it is preferable to use the function $1+\xi(r)$.

VII. Discussion and Perspectives

In this Chapter we have introduced a modification of the β model of turbulence which may be used to describe the large scale galaxy distribution. The model generates a smooth transition from small-scale self-similarity to large-scale homogeneity and it is associated with a scale-dependent fractal behavior. Due to the smooth transition from fractality to homogeneity, the fractal dimension of the galaxy distribution is 'dressed' from an asymptotic value $D_{as}=1$ (suggested by dimensional arguments) to an observed, 'effective' value D_{ef}. Distributions which closely resemble the real galaxy catalogs are associated with a dressing of the fractal dimension from the value $D_{as}=1$ to a value $D_{ef} \approx 1.2$.

The phyiscal interpretation of the β model has been given in the context of both CDM and HDM scenarios for galaxy formation. In the CDM scenario, the small-scale region with dimension $D=1$ is associated with the scales which have undergone 'fully developed' gravitational clustering, while the large scales still retain the homogeneous character of the initial density field. The transition from one behavior to the other is presumably found at the scales which are now entering the nonlinear regime. In this context, the transition region has a scale-dependent nature, being shifted to larger and larger scales as the gravitationally clustered region grows. In a HDM scenario, the cascading β model can be physically rooted in the process of pancake fragmentation; the scale-dependent survival probability may be associated in this case with a scale-dependent balance between gravitational clustering and hydrodynamical effects.

Another approach to the use of the scale-dependent β model, however, is based on considering this model as a 'fractal machine', capable of producing point distributions with controllable fractal properties. It is thus possible to test data analysis methods and to verify the appropriateness of inferences that might be difficult to evaluate in the analysis of real galaxy catalogs. In this sense, the β model discussed here may be considered as a 'reference point' in the study of the fractal behavior of the large scale distribution of matter. For example, we are now considering the comparison between the statistical properties of the real galaxy distributions and those generated by the cascading β model, in order to obtain the 'best' form of the survivival probability $p(l_k)$. Solving this 'inverse problem' could perhaps provide a useful framework for the description of the large scale galaxy distribution.

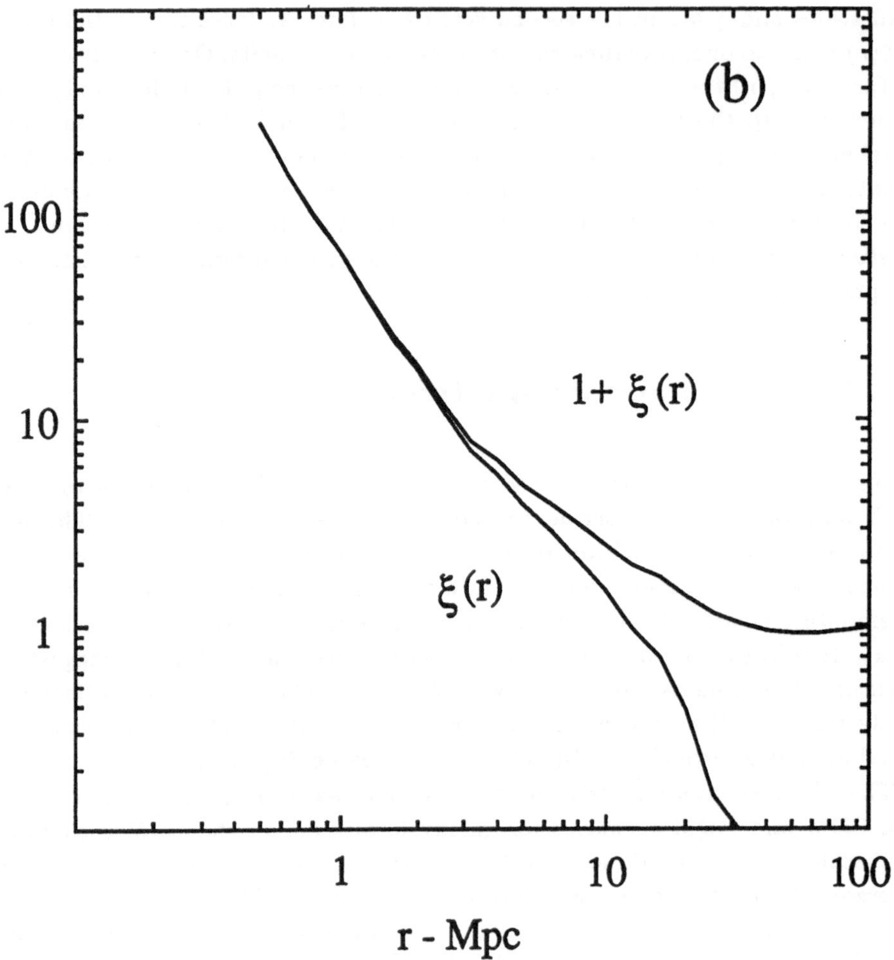

Fig. 4b. The corresponding two-point correlation functions $1+\xi(r)$ and $\xi(r)$. The length scales are only indicative.

Along this line of reasoning, we have considered a 'staircase' β model in which the 'characteristic' dimension varies from D=3 at large scales, to D=2 at intermediate scales and finally to D=1 at small scales. This model has been suggested by the recent analysis of the Perseus-Pisces region redshift data (Guzzo et al. 1990), where evidence of two scaling regimes with dimension $D \approx 1$ and $D \approx 2$ has been found. These results are in accordance with an observation by Luckas and Novikov (1988), who have recently suggested that the fractal dimension of the galaxy distribution may have a physically motivated scale-dependence, growing from $D \approx 1$ at small scales, where a filamentary structure dominates, to $D \approx 2$ at large scales, where pancakes dominate (see e.g. Shandarin and Zeldovich 1989), to $D \approx 3$ at very large scales, where homogeneity and space-filling distributions are finally found. The use of the 'staircase' β model has quantitatively confirmed the observation by Guzzo et al. (1990) that the function $1 + \xi(r)$ is the appropriate quantity to be used when studying the fractal behavior of the galaxy distribution.

A simplified assumption of the cascading β models discussed above is that they consider only 'active' or 'dead' objects, the latter having a density which is too low to generate visible objects. This leads to the existence of large voids which are completely devoid of galaxies and to a strictly mono-fractal behavior. A physically relevant extension of the 'staircase' β model is based on considering a continuous redistribution of matter instead of a yes/no behavior. A detailed discussion on this type of models will be given in a forthcoming paper. Here we only note that in these new models there are two basic parameters, namely the 'survival' probability $p(l_k)$ and the redistribution factor $f(l_k)$. The quantity $p(l_k)$ must now be interpreted as the probability of a positive density fluctuation (i.e. the probability of a son object to have a density larger than the parent object), while the variable $f(l_k)$ measures the ratio of the density ρ_{under} of the underdense 'sons' to the density ρ_{parent} of the parent object, i.e. $\rho_{under} = f(l_k)\rho_{parent}$. The density of the overdense regions is then fixed by conservation of mass. The quantity $f(l_k)$ may be either a deterministic function or a random variable. Physically, the quantity $p(l_k)$ varies from p=1/2 at large scales to p=1/4 at small scales, where nonlinear gravitational clustering generates a fractal dimension D_{a_s}=1. A large-scale probability p=1/2 has a natural explanation if in the field of the initial, linear random density perturbations positive and negative fluctuations around the mean density are equally probable. The transition regions is fixed at the scales which are just entering the nonlinear regime. Larger scales are still associated with linear density fluctuations. The quantity $f(l_k)$, on the other hand, varies from a value f=1 at large scales, corresponding to homogeneity (for f=1 the density of the sons is equal to the density of the parent object), to a value f=0 at small scales. The value f=0 corresponds to a fully clustered regime; this value is typical of the β model. In fact, in the β model the density of the underdense regions is zero. When f is different from zero, a multifractal behavior can be generated, since the process is multiplicative. When f=0, the cascade becomes mono-fractal. Consequently, in these models a smooth transition from multifractality at large scales to self-similarity at small scales is generated, together with a natural scale-dependence of the fractal behavior. The large-scale homogeneity is generated by the scale-dependence of the parameter $f(l_k)$, which is in turn associated with the spectrum of linear density perturbations. These models are thus capable of producing a density field $\delta\rho/\rho$ which has scale-dependent fractal properties, with a peaked and irregular behavior which is more pronounced at small scales. Individual objects such as single galaxies are associated with density peaks which are higher than a predetermined threshold density. The time evolution is then reflected by the growth of the gravitationally clustered scales, i.e. by a displacement of the transition region from p=1/4 to p=1/2 to larger and larger scales.

Fig. 5 shows a slice of a three-dimensional distribution generated by the above model for an exponential dependence of $f(l_k)$ and for a threshold density $\rho_{tr} = 500\rho_0$ where ρ_0 is the average density at large scales. There are rich 'Abell' clusters and filaments as well as isolated galaxies; the voids are not completely empty and

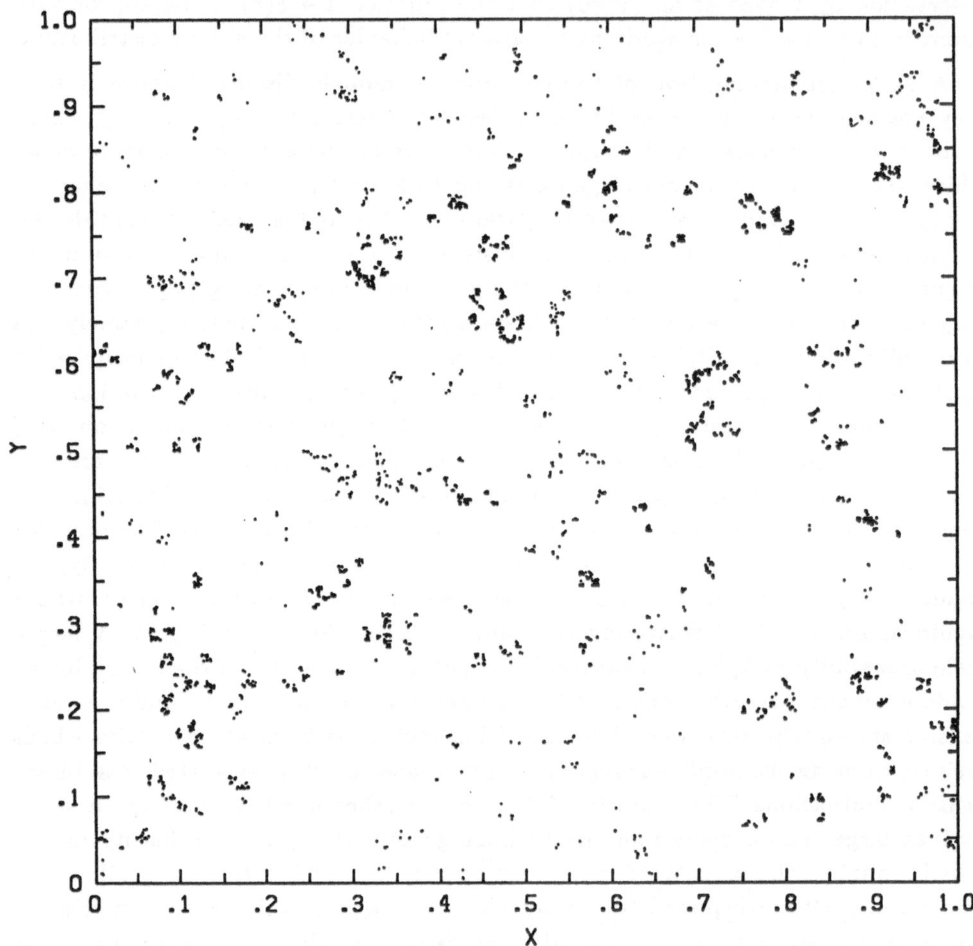

Fig. 5. Two-dimensional slice of a three-dimensional distribution generated by the multifractal version of the β model introduced in Sect. VII. The points represent individual galaxies and correspond to peaks in the density field which are at least 500 times larger than the average density at large scales.

in general the distribution bears several close resemblances with the real galaxy distribution. Further details on this model and on its statistical properties are given in a forthcoming paper (Castagnoli et al, in preparation).

Obviously, the cascading models described above are over-simplifyed descriptions of the physical processes which determine the large scale structure of the Universe. However, we believe that the search for 'minimal' fractal models based upon well-defined physical mechanisms may shed some additional light on the properties of the real Universe.

Acknowledgements

A large portion of this work has been done in collaboration with Prof. C. Castagnoli, who stimulated my interest in this subject. I am grateful to A. Blanchard, S. Borgani, U. Frisch, L. Guzzo, P. Galeotti and R. Valdarnini for useful discussion and comments. Part of the numerical codes used in this work have been developed by G. Murante.

References

Bachall, N. (1988): *Ann. Rev. Astron. Astrophys.* **26**, 631.

Barenblatt, G. (1980): *Similarity, Self-Similarity and Intermediate Asymptotics* (Plenum Press, New York).

Blanchard, A., Alimi, J.M. (1988): *Astron. Astrophys.* **203**, L1.

Calzetti, D., Giavalisco, M., Ruffini, R. (1988): *Astron. Astrophys.* **198**, 1.

Castagnoli, C., Provenzale, A. (1990): in *Fractals in Astronomy*, Ed. A. Heck, *Vistas Astron.* **33**, 323.

Castagnoli, C., Provenzale, A. (1991): *From Small-Scale Fractality to Large-Scale Homogeneity: A Family of Cascading Models for the Galaxy Distribution*, *Astron. Astrophys.*, in press.

Coleman, P.H., Pietronero, L., Sanders, R.H. (1988): *Astron. Astrophys.* **200**, L32.

Davis, M., Peebles, P.J.E. (1983): *Astrophys. J.* **267**, 465.

Fournier d'Albe, E.E. (1907): *Two New Worlds: I. The Infra World; II. The Supra World* (Longmans Green, London).

Frisch, U., Sulem, P., Nelkin, M. (1978): *J. Fluid Mech.* **87**, 719.

Grassberger, P., Procaccia, I. (1983): *Physica* **9D**, 189.

Guzzo, L., Iovino, A., Chincarini, G. (1990): in *Physical Cosmology* (Frontières, Paris).

Hoyle, F. (1953): *Astrophys. J.* **118**, 513.

Jones, B.J.T., Martinez, V.J., Saar, E., Einasto, J. (1988): *Astrophys. J. Letters* **332**, L1.

Liu, Y.Z., Deng, Z.G. (1988): in *High-Energy Astrophysics*, Ed. G. Borner (Springer-Verlag, Berlin), p. 205.

Lukash, V.N., Novikov, I.D. (1988): *Acad. Sciences USSR - IKI*, preprint.

Mandelbrot, B.B. (1982): *The Fractal Geometry of Nature* (Freeman, San Francisco)

Martinez, V.J. (1990): in *Fractals in Astronomy*, Ed. A. Heck, *Vistas in Astronomy* **33**, 337.

Martinez, V.J., Jones, B.J.T. (1990): *Monthly Not. Roy. Astron. Soc.* **242**, 517.

Martinez, V.J., Jones, B.J.T., Dominguez-Tenreiro, R., van de Weygaert, R. (1990): *Astrophys. J.* **357**, 50.

Paladin, G., Vulpiani, A. (1987): *Phys. Rep.* **156**, 147.

Peebles, P.J.E. (1980): *The Large-Scale Structure of the Universe* (Princeton Univ. Press, Princeton).

Peebles, P.J.E. (1988): preprint.

Peebles, P.J.E. (1990): in *Fractals in Physics*, Eds. A. Aharony, J. Feder (North Holland, Amsterdam).

Pietronero, L. (1987): *Physica* **144A**, 257.

Provenzale, A. (1990), in *Physical Cosmology* (Frontières, Paris).

Rivolo, A.R. (1986): *Astrophys. J.*, **301**, 70.

Ruffini, R., Song, D.J., Taraglio, S. (1988): *Astron. Astrophys.* **190**, 1.
Schramm, D. (1988): in *Gauge Theory and the Early Universe*, Eds. P. Galeotti, D. Schramm (Kluwer Acad. Publ., Dordrecht).
Schulman, L.S., Seiden, P.E. (1986): *Astrophys. J.* **311**, 1.
Shandarin, S.F., Doroskhevic, A.G., Zeldovich, Ya.B. (1983): *Sov. Phys. Usp.* **26**, 46.
Shandarin, S.F., Zeldovich, Ya.B. (1989): *Rev. Mod. Phys.* **61**, 185.
Soneira, M.S., Peebles, P.J.E. (1978): *Astron. J.* **83**, 845.
Turok, N. (1986): in *Cosmology, Astronomy and Fundamental Physics*, Eds. G. Setti, L. Van Hove, *ESO Conference and Workshop Proc.* **23**, p. 175.
Valdarnini, R., Borgani, S., Provenzale, A. (1991): *Fractal Properties of Cosmological N–Body Simulations*, submitted.
Vicsek, T., Szalay, A.S. (1987): *Phys. Rev. Letters* **58**, 2818.
White, S.D.M. (1987): in *Nearly Normal Galaxies*, Ed. S. Faber (Springer–Verlag, Berlin).

Is the Spatial Distribution of Galaxies a Bounded Self–Similar Structure? Observational Evidences

Daniela Calzetti [1,2], Mauro Giavalisco [1,3]

[1]on leave from: I.C.R.A. – Dipartimento di Fisica, Università di Roma, Piazzale Aldo Moro 2, I–00185 Roma, Italy
[2]Space Telescope Science Institute, 3700 San Martin Drive, Baltimore, MD 21218 – U.S.A.
[3]Dept. of Earth and Space Science, State University of New York, Stony Brook, NY 11794–2100 – U.S.A

Abstract: We want to show here that the distribution of galaxies in space may be described in terms of a self-similar arrangement of the clustering on scales of distances ranging from some Mpc to $(30-40)h^{-1}$ Mpc. To this end, the classical two-point correlation function has been evaluated both in the angular and the spatial case on the Zwicky and Cfa1 catalogues, respectively. In contrast to previous works, the quantity $1+w(\theta)$ and its spatial equivalent $1+\xi(r)$ have been analysed with respect to their scaling properties with the sample depth. We have not studied the local scaling properties of the distribution (multifractality) and have approximated it to a simple fractal set. The resulting estimate for the fractal dimension (or better, the correlation dimension) is $D_2 \approx 1.7$ from angular data, while $D_2 \approx 2.3$ from the spatial ones. If this value is confirmed, the clustering in the galaxy distribution results to be less strong, but more extended in space, than previously found.

1. Introduction

The discovery of the existence and the homogeneity of the blackbody microwave Cosmic Background Radiation and the matching of observations with the estimated abundance of light elements coming from nucleosynthesis have constituted some of the most important observational evidences in favour of the Hot Big Bang cosmology (Gamow 1948), based on the FRW metric. According to this cosmological model the Universe must have been a physical system characterized in the past by a highly homogeneous and isotropic distribution of matter and energy in space.

Neverthless, it is evident that at the present this expected homogeneity is not actually observed in the distribution of the astrophysical systems, at least over scales of distances covered by typical cosmological surveys. This means that the processes that have led to the formation of structures have also created the observed irregularity and the Friedmannian regularity in the distribution of matter in the Universe is a fact which is restored only if an average of the density is considered over appropriate volumes of space. The question under discussion is the size up to which the presence of structures gives rise to inhomogeneity, but it is clear that its value has constantly increased as deeper red-shift catalogues have been available. In fact, the detection of regions void of objects, be they galaxies or clusters, and of large clumpy structures (to give some references: Chincarini and Rood 1975, 1976; Chincarini 1978; Tarenghi et al. 1980; Einasto et al. 1980; Kirshner et al. 1981; Chincarini et al. 1981; Bahcall and Soneira 1982a,b; for a review, see Oort 1983; De Lapparent et al. 1986; Haynes and Giovanelli 1986; Chincarini and Vettolani 1987) has shown that the scale over which inhomogeneity is still detectable may be as large as $\sim 100h^{-1}$ Mpc ($H_o = 100km/s/Mpc$ is assumed through this work).

The explanation of the existence of such large structures is one of the main tasks of the theories of galaxy formation. To this end, the distribution of galaxies, clusters and superclusters has been intensively studied on the available catalogues (to quote some: the Zwicky catalogue by Zwicky et al. 1961-68; the Lick catalogue by Shane and Wirtanen 1967; the Abell catalogue by Abell 1958; the CfA catalogue by Huchra et al. 1983-..; the IRAS catalogue by Lonsdale et al. 1989; the ACO catalogue by Abell et al. 1989) in order to search, within the evident 'disorder', for possible statistical properties, which can describe on average the structures. In this respect, since the work mainly of Peebles and his collaborators during the 70's, the correlation function of the density field traced by luminous matter has been particularly popular. In principle, the knowledge of all the orders of correlation allow to describe completely the structure analyzed (Peebles 1980), but the determination of orders higher than two is difficult for computational problems and for increasing indetermination in the results. Anyway, at the price of missing the exact picture of the distribution, it is possible to draw interesting conclusions about the clustering in the Universe from a very simple way of investigation, namely the two-point correlation function. Because of its simplicity in use, although a number of alternative statistical methods have been also developed (as for instance the void probability function, see White 1979), none has reached the same popularity as the correlation function.

The two-point correlation function $\xi(r)$ gives the excess-from-random probability of finding an object in an elementary volume $d\,V_1$ and, at the same time, another object in $d\,V_2$ at distance r from the first (Peebles 1980):

$$d\,P_{12} = <n>^2 [1 + \xi(r)] d\,V_1\,d\,V_2, \qquad (1)$$

where $d\,P_{12}$ is the (unnormalized) joint probability and $<n>$ is the average density in our sample.

For the two–dimensional sphere, instead of the tridimensional space, an analogous expression can be written:

$$d\,\mathcal{P}_{12} = \mathcal{N}^2[1 + w(\theta)]d\,\Omega_1\,d\,\Omega_2, \tag{2}$$

where $\mathcal{N}$ is the surface density, θ is the angular distance between the two objects as seen by us, $d\,\Omega_{1,2}$ are the elementary solid angles (Peebles 1980).

Usually the correlation functions ξ or w are actually used in observational work and the data are fitted to some model for them. We will see, however that it is safer not to work with the excess of probability, but with the probability itself, that is with $1 + \xi$. This is necessary if one wants to analyze the scaling properties of the distribution (Pietronero 1987), but it is also the case when the correlation becomes particularly small and one is trying to fit some model, especially when working with logarithms. We are going to show that present optical data about galaxy distribution, properly analyzed by the two-point correlations (both angular and spatial), show that the distribution itself exhibits self-similar or fractal features on scales starting from some Mpc up to an upper cut-off of about 35 - 40 h^{-1} Mpc after which a transition region leading to homogeneity should start. At the moment, we have no physical explanation for this, but it is clear that it is an issue that must be considered by theories about galaxy formation and it is possible that its existence is due to a particular physical process which might have played an important role in the field.

2. Observational Results for a Self-Similar Structure

The two-point correlation function of galaxies and clusters has been and presently remains widely studied, in order to extract as much information as possible from the distribution of objects. The main results, coming either from bidimensional and tridimensional catalogues, can be summarized as follows:

1. the galaxy correlation function behaves as an inverse power law:

$$\xi_g(r) \approx \left(\frac{r_{0_g}}{r}\right)^{\gamma_g} \qquad \gamma_g \simeq 1.77 \tag{3}$$

   where the correlation length $r_{0_g} \simeq 5h^{-1}$ Mpc (Peebles and Hauser 1974; Peebles and Groth 1975; Groth and Peebles 1977; Davis and Peebles 1983; Bean et al. 1983).

2. Clusters also have a correlation function of shape:

$$\xi_c(r) \approx \left(\frac{r_{0_c}}{r}\right)^{\gamma_c} \qquad \gamma_c \simeq 1.8, \tag{4}$$

   with r_{0_c} ranging from about 25 to $50h^{-1}$ Mpc, according to the richness of the cluster (Bahcall and Soneira 1983; Klypin and Kopylov 1983; Bahcall et al. 1988; Batuski et al. 1989; for a wide review on the topic, see Bahcall 1988).

3. analyzing the galaxy correlation length, Einasto et al. (1986) have found:

$$\log r_o = \log R + const, \tag{5}$$

where R is the depth of the sample considered. This result has been recently criticized by Blanchard and Alimi (1988), but other authors (Davis et al. 1988 and references therein) have obtained a correlation length depending on the sample size on both the Northern and the Southern Hemisphere, so we will consider the above Eq. (5) an indication that the correlation length is not a well–defined quantity.

Since $\xi(r)$ is related to the number density behaviour $n(r)$ through the expression (Peebles 1980):

$$n(r) = < n > [1 + \xi(r)], \tag{6}$$

it is evident that Eqs.(3), (4) and (5) imply: a) that the intermediate scale structure is highly inhomogeneous; b) that the density follows an inverse power law; c) that this behaviour is common to different clustering hierachies (galaxies and clusters), since $\gamma_g \simeq \gamma_c$; d) that there is no preferred clustering scale, because the correlation length is not defined (see also Coleman et al. 1988). All these considerations lead to suppose that the intermediate scale distribution ($\lesssim 100h^{-1}$ Mpc) of galaxies, clusters, and so on, in the Universe can be modeled by some kind by a self-similar structure (Mandelbrot 1975, 1979, 1983; Pietronero 1987; Calzetti et al. 1987).

We want to stress that the increase of the correlation length has been interpreted also in two other ways differing from the fractal one: a) variation of the number density of the galaxies due to the fact that the samples are not still fair (Börner and Mo 1990); b) dependence of the galaxy clustering on some particular features of galaxies as morphology and luminosity (Davis et al. 1988; Börner and Mo 1990).

In addition, another argument against fractality has been the scaling of the amplitude of the angular correlation function $w(\theta)$ with the sample depth (Groth and Peebles 1977; Peebles 1980, §62).

We see in the following that a different choice in the way of dealing with the two-point correlation function gives the possibility of explaining some observational evidences in terms of a self-similar structure with an upper cut-off R_{co}, solving also the problem of the $w(\theta)$-amplitudes. The scale R_{co} marks the transition towards homogeneity, essential for recovering the homogeneous and isotropic large–scale Universe (Ruffini et al. 1988).

3. The Model

The most general self-similar structure is the multifractal, characterized by a continuous spectrum of generalized fractal dimensions (Mandelbrot 1974; Frisch and Parisi 1985; Halsey et al. 1986. For observational reports about multifractality in galaxy distribution, see Atmanspacher et al. 1989; Martinez and Jones 1990 and references therein). A simple fractal may be considered as a particular multifractal with a degenerate spectrum of dimensions. By the way, working with the two-point correlation functions, only one of the infinite dimensions appears from the data, i.e., the 'correlation dimension' D_2. Then it is equivalent to deal with a simple fractal with dimension:

$$D = D_2 = 3 - \gamma \qquad (7)$$

and this will be assumed in the work. But it is clear that by fractal we mean a general self-similar distribution, like a multifractal. This because we are, by now, interested in exploring the scale range in the Universe in which the self-similarity is exhibited.

A fractal is a highly inhomogeneous structure, where some 'regularities' can be recovered if the 'conditional cosmological principle' is followed (Mandelbrot 1983): only the physical points of the set can be considered observers. In this case, the distribution is characterized by a number density behaviour:

$$n(r) = k\, r^{-\gamma}. \qquad (8)$$

Introducing an upper cut-off R_{co}, we can specify:
for $R_s < R_{co}$:

$$n(r) = \frac{3-\gamma}{3} R_s^{\gamma} < n > r^{-\gamma}; \qquad (9)$$

for $R_s > R_{co}$:

$$n(r) = \begin{cases} \frac{3-\gamma}{3} R_{co}^{\gamma} < n > r^{-\gamma} & r \leq R_{co} \\ < n > & r > R_{co} \end{cases}, \qquad (10)$$

where, R_s is the sample depth and $< n >$ is the asymptotic average density (obtained for $R_s \gg R_{co}$). Actually, the transition occurring at $r = R_{co}$ describes a very crude model (Calzetti et al. 1989), since the density presents here a discontinuity. Such a model is, however, assumed as a first approximation of a more realistic one and its simplicity makes calculations handy.

For the case $R_s < R_{co}$, Eq. (6) becomes:

$$1 + \xi(r) = A\, r^{-\gamma} \qquad (11)$$

with:

$$A = \frac{3-\gamma}{3} R_s^{\gamma} \qquad (12)$$

(Calzetti et al. 1988), implying that $1 + \xi(r)$, and not $\xi(r)$, behaves in this model as a pure power law and that, for distances smaller than the fractal size R_{co}, the correlation length depends on the sample depth. Although, for $1 + \xi(r) \gg 1$, $1 + \xi(r) \approx \xi(r)$ and the two quantities have exactly the same physical content, neverthless working in double logarithmic scales for getting their paramenters, as usually done, gives different information.

The angular two-point correlation function is the projection on a sphere of the spatial correlation function (Limber 1953; Peebles 1980). In general, the most popular catalogues for galaxies are the apparent magnitude limited ones, so $w(\theta)$ is function of the equivalent depth of the catalogue R_* (Groth and Peebles 1977), defined implicitly as:

$$M_* \simeq m - 25 - 5\log[R_*(1+z)] - \kappa z,$$

where M_* is the absolute magnitude parameter of the adopted luminosity function, m is the apparent magnitude limit of the catalogue, z is the red-shift corresponding to R_* and κz is the so-called κ-correction. This definition does not account for the possibility that the density distribution may be clustered, but the approximation is sufficient for a large number of applications (Calzetti, Giavalisco, Ruffini, 1990, in progress).

Remembering that $\beta = \gamma - 1$ and that we are working in the small angular approximation $\theta \ll 1$, we get:
a) when $R_{co} \ll R_*$:

$$1 + w(\theta) = B(R_*, \gamma)\theta^{-\beta}, \qquad B(R_*, \gamma) \propto R_*^{-\gamma} \tag{13}$$

(homogeneous scenario; see Groth and Peebles 1977; Peebles 1980).
b) for $R_{co} \gg R_*$:

$$1 + w(\theta) = B\theta^{-\beta}, \qquad B = const \tag{14}$$

(indefinite fractal scenario; see Mandelbrot 1975; Peebles 1980).
c) when $R_{co} \simeq R_*$:

$$1 + w(\theta) \simeq B(\gamma, R_{co}/R_*)\, \theta^{-\beta}, \tag{15}$$

where:

$$B(\gamma, x) \propto x^{\gamma} \frac{\frac{3-\gamma}{3}x^{\gamma}D_1 + D_2}{\left(\frac{3-\gamma}{3}x^{\gamma}M_1 + M_2\right)^2}, \tag{16}$$

with D_1, D_2, M_1, M_2 containing the integrals of the galaxy luminosity function (Calzetti et al. 1989).

In order to compare the self-similar model with the data from observations, we can adopt two alternatives:
1) to work with the angular correlation function and verify that the behaviour of $1 + w(\theta)$ agrees with a power law. Then, estimate β and R_{co} from the trend, respectively, of $1 + w(\theta)$ and of the amplitude B with the equivalent depth R_* (Calzetti et al. 1990).
2) to deal with the spatial correlation function; again, verify the power law behaviour of $1+\xi(r)$. Subsequently, estimate the value of γ and check the agreement of the amplitude A with the law in Eq. (12).
If both results are available, a cross-checking between the results is essential to test their consistency.

4. Determination of the Angular Correlation Function from the Zwicky Catalogue

The angular two-point correlation function is the simplest one to determine, since it needs no information about the galaxy red-shift, avoiding thus the problems related to estimation of distances, selection functions and, at small angular separations, boundary effects (Davis and Peebles 1983). The only connection with distances is given by the equivalent depth R_*, which, in any case, is not needed with high accuracy (Calzetti et al. 1990).

In order to estimate $1+w(\theta)$ and, eventually, to verify the self-similar behaviour of the galaxy distribution, the Zwicky catalogue (Huchra 1987) has been analyzed (Calzetti et al. 1990). The catalogue is complete up to $m = 15.5$ and for $\delta > 0^o$. To minimize the galactic obscuration effects, only galaxies with $b > 40^o$ has been considered.

The catalogue has then been cut in five apparent magnitude limited sub-samples, starting from $m = 13.5$ and increasing with step $\Delta m = 0.5$. Residual obscuration has been studied and removed using the correction (Bahcall and Soneira 1983):

$$P(b) = a\ 10^{d(1-csc|b|)}. \qquad (17).$$

The values of the amplitudes B and of the index β for each sub-sample have been estimated following the standard procedure (Peebles, 1980).

One question we have to answer is whether it is $1+w(\theta)$ or $w(\theta)$ that matches better a straight line with the same slope for *all the sub-samples*. From Fig. 1 it is evident that the more stable values of β are given by $1+w(\theta)$.

The fact that in this case $\beta \simeq 0.3$, corresponding to $\gamma \simeq 1.3$, less than the usual quoted value ~ 1.8, is not a novelty, since similar values for γ have been previously obtained by other authors (de Lapparent et al. 1988; Wiedenmann and Atmanspacher 1990; Babul and Postman 1990). We do not compare our β with the slopes of $w(\theta)$ obtained in other papers (as, i.e., the one by Babul and Postman 1990), because, in contrast with $\xi(r)$, the angular correlation function is of the order of unity in amplitude and estimating $1+w(\theta)$, instead of $w(\theta)$, is critical for the slopes.

A fit of the amplitudes B (see Fig. 2) enables us to calculate the value of R_{co}, imposing the previously estimated value of γ. The cut-off size resulting from the data is $(32\pm6)h^{-1}$ Mpc (Calzetti et al., 1990), higher than the value $\sim 10h^{-1}$ Mpc determined some years ago (Peebles and Groth 1975; Groth and Peebles 1977) and in line with recent results (Maddox et al. 1990).

One of the problems connected with the estimate of the amplitude B is the presence of the luminosity function in its expression Eq. (16) (for an extensive review on this topic, see Binggeli et al. 1988), which is not uniquely determined and, although it appears in Eq. (16) only through its integral, different shapes may, in principle, lead to different results. This effect has been tested with two luminosity functions: the Abell one (Abell 1962) and the Schecter one (Schecter 1976), this last with the parameters given in de Lapparent et al. (1989). The

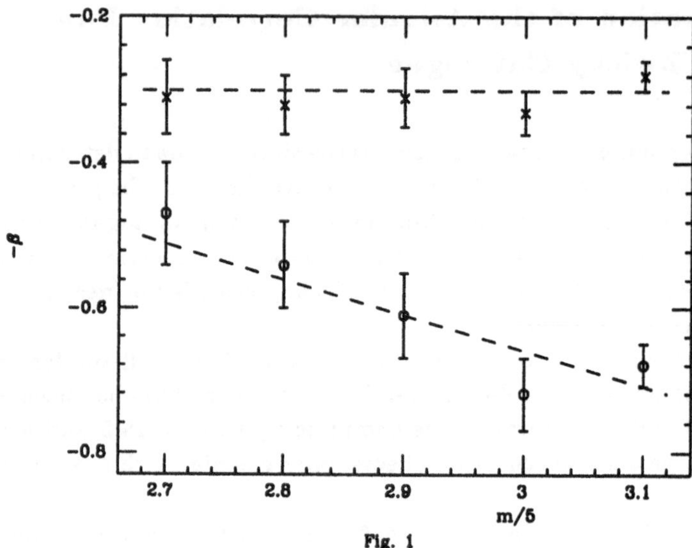

Fig. 1. The slopes β, with their 1σ uncertainties, are shown as function of $m/5 \propto \log R_*$, for both cases of $1+w(\theta)$ (crosses) and $w(\theta)$ (circles). It is evident that β_{1+w} is compatible with a constant line (drawn for comparison), while β_w is not constant and the fit gives a slope $\simeq -0.8$ (from Calzetti et al., 1991).

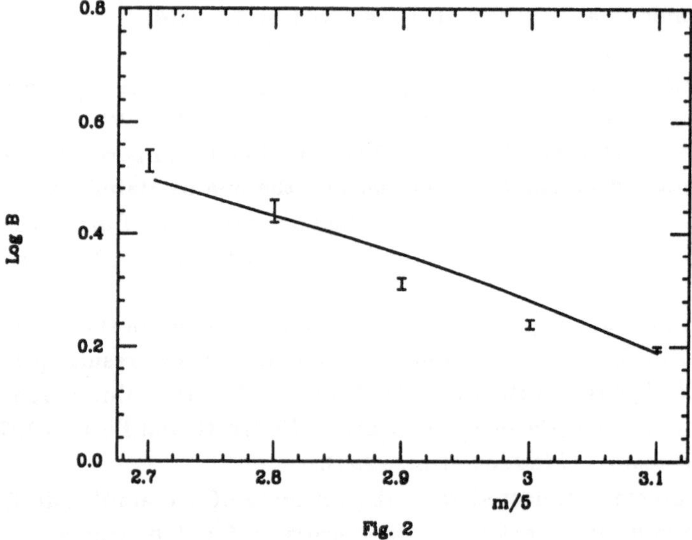

Fig. 2. The amplitudes B of $1 + w(\theta)$ for the five Zwicky sub-samples. The line is the best fit obtained imposing $\gamma = 1.3$ (from Calzetti et al. 1990).

final results are compatible within the uncertainties on the data. There is another possible shape for the luminosity function: the one by Lilje and Yahil (1988, quoted and with parameters determined in Lahav et al. 1988), which corrects the bright branch of the Schecter function. We are confident that a correction in the less populated part of the luminosity distribution is not critical for our conclusions.

5. The Spatial Correlation Function Estimated from the CfA1 Catalogue

In order to further investigate the self–similar correlation properties of the galaxy distribution and to get a cross check with the results of the angular analysis we have studied the spatial correlation function and its scaling properties. To this purpose we have used the CfA1 catalogue in which redshifts are available for galaxies brighter than the 14.5^m. The catalogue is complete up to this magnitude. This survey covers the same sky surface than the Zwicky catalogue, but clearly it is less deep and it is contained within it. The above sample was then divided into 16 sub-samples of different apparent magnitude limits, starting from $m_c = 13.0^m$ to $m_c = 14.5^m$ with increments of $\delta m_c = 0.1^m$.

Given as usual the equivalent depth of the catalogue R_* (Groth and Peebles 1977), which allows to approximate the observational sample with a cone of height R_* and with the vertex of the observer, the 'equivalent sample radius' R_s is then defined as the radius of the sphere with the same volume as the cone:

$$R_s = \left[\frac{3\,(1 - cos\theta)}{2}\right]^{1/3} R_*, \tag{18}$$

where in our case $\theta \approx 50^o$.

The correlation function has been estimated by:

$$1 + \xi(r) = \frac{N_{oo}(r)}{N_{or}(r)}, \tag{19}$$

where $N_{oo}(r)$ is twice the number of pairs of galaxies having separation in the range dr at r and $N_{or}(r)$ is an analogous quantity, but now the first galaxy of each pair always belongs to the observational sample while the second belongs to a properly generated random sample (see Davis and Peebles 1983). Actually more than a sample is actually generated in order to reduce the noise due to the fluctuations and N_{or} is an average over all the random catalogues.

In Figs. 3a-3d we have plotted the correlation function $\xi(r)$ versus the separation r in double logarithmic scale for the 13.0^m, 13.5^m, 14.0^m and 14.5^m sample. In Figs. 4a-4d the quantity $1 + \xi(r)$ is plotted as well.

We may at first note that $1 + \xi$ keeps showing the features of a power law function for separation ranges larger than in the case of ξ. This suggests that the break observed in the latter plot at $r \approx 10, 13$ Mpc may be not physical, but due to the logarithmic divergence when ξ approaches zero. This fact may be

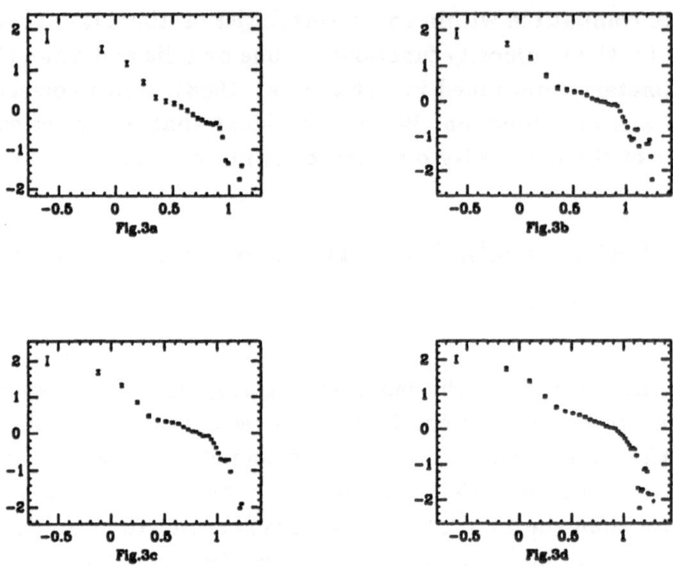

Fig. 3. The function $\xi(r)$ as measured from the 13.0^m, 13.5^m, 14.0^m and 14.5^m subsamples plotted versus the separation r in double logarithmic scale. Note the apparent breaking down from the power law from separation of the order of 10 Mpc.

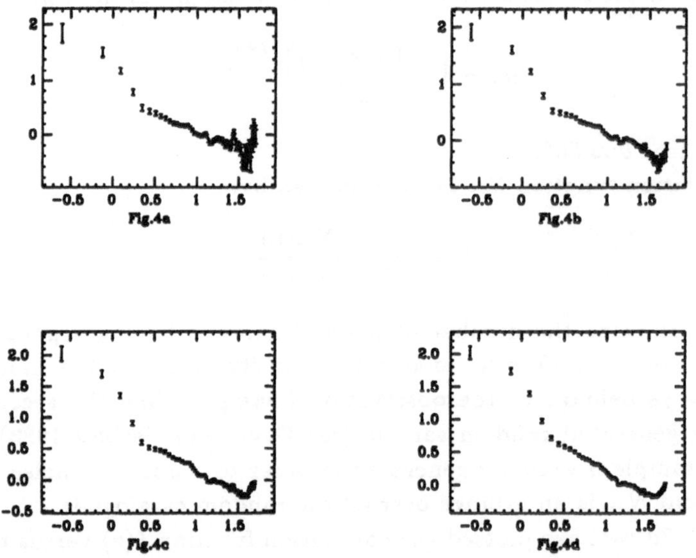

Fig. 4. The function $1 + \xi(r)$ as measured from the 13.0^m, 13.5^m, 14.0^m and 14.5^m subsamples plotted versus the separation r in double logarithmic scale. Note how the power law behaviour is now much more extended than in the case of $\xi(r)$.

misleading and, as in the angular case, we prefer to conduct our analysis using $1+\xi$ as a safer indicator of the correlation. Actually, as can be seen both from Figs. 3 and 4, the correlation function exhibits a double power law behaviour. The separation value r_t at which the transition occurs is of the order of 3 Mpc. For $r > r_t$ (the intermediate scale regime) and up to $r \sim 40$ Mpc $1+\xi$ seems to show a unique trend, also if some distorsions are present, especially in the smaller samples. However, when fitting data we restrict ourselves to the interval (3; 20) Mpc.

It may be argued that observations have shown that the correlation function cannot be described as a unique power law, as that in Eq. (12) and the comparison may be then misleading. New scaling relations are then necessary to model the data. Following the observational evidence, we can assume that $1+\xi_1(r) = A_1 r^{-\gamma_1}$ for $r < r_t$ and $1+\xi_2(r) = A_2 r^{-\gamma_2}$ for $r > r_t$. Imposing the continuity for $r = r_t$ and using the normalization condition $\int_0^{R_s} \xi(r) r^2 dr = 0$ (Calzetti et al. 1988), it is easy to find the scaling relation for the two amplitudes:

$$A_1 = \frac{3-\gamma_2}{3} R_s^{\gamma_2} \frac{r_t^{\gamma_1-\gamma_2}}{\left(\frac{r_t}{R_s}\right)^{3-\gamma_2}\left(\frac{3-\gamma_2}{3-\gamma_1}-1\right)+1} \sim \frac{3-\gamma_2}{3} R_s^{\gamma_2} r_t^{\gamma_1-\gamma_2}, \tag{20}$$

$$A_2 = \frac{3-\gamma_2}{3} R_s^{\gamma_2} \frac{1}{\left(\frac{r_t}{R_s}\right)^{3-\gamma_2}\left(\frac{3-\gamma_2}{3-\gamma_1}-1\right)+1} \sim \frac{3-\gamma_2}{3} R_s^{\gamma_2}. \tag{21}$$

Indeed we have observed that $\gamma_1 \approx 1.9$ (see also Guzzo et al., 1991) and $\gamma_2 \approx 0.7$ and that the lowest value for R_s is $R_s \sim 6\,r_t$, so the fractional part in the above equations is always of the order of 1. We see that the correlation function for the intermediate scale regime continues to scale as a simple self-similar distribution with correlation dimension $D_2 = 3-\gamma_2$. In the small scale regime the correlation function has an extra term due to the existence of the transition, but it is important to observe that in contrast to the previous case the scaling law has a slope equal to γ_2 as well, and not equal to the slope of each correlation function, that is γ_1.

In Figs. 5a-5b we show the correlation amplitude for each subsample as well as their slopes as a function of R_s for the case of $1+\xi$. Remarkably, we observe a power law scaling of the correlation amplitude with the sample depth continuing up to the deepest sample where the equivalent sample radius is about 35 Mpc. A linear fit produces the slope $\gamma_A = 1.0 \pm 0.1$. For what concerns the correlation slopes we may see thay they have an increasing trend with R_s. Probably edge effects may be partially responsible for this fact, but a correlation between the parameters of the power law when fitting data is certainly present, particularly if an expression like Eq. (12) holds. We are actually working on this problem. It is, however, interesting to observe that in the angular correlation analyses this problem seems to be less serious. For the moment we can use our simple statistics to compare with the other results and in the following we will take for the correlation slope the average value over the 16 subsamples, that is $\bar{\gamma} = 0.7$. We want also explicitly to observe that the results for ξ in the 14.5^m sample, i.e. the slope and the amplitude, are quite in agreement with what has been reported in previous works (de Lapparent et al. 1988; Babul and Postmann 1990). It is interesting to observe the discrepancy between the slopes obtained from the angular and spatial correlations. The first is $\gamma_{an} \approx 1.3$, while the second $\gamma_{sp} \approx 0.8$. If we remember that the spatial correlation is actually evaluated in the redshift space rather than in the physical space, an explanation of the discrepancy is to be searched in the properties of the peculiar velocity field of galaxies.

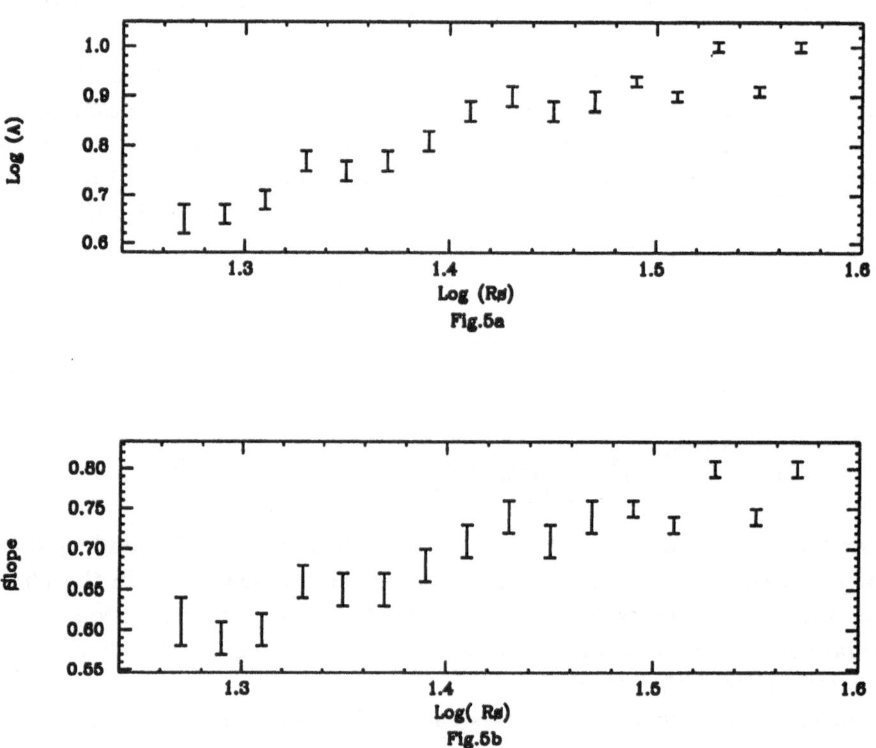

Fig. 5. *a)* the spatial correlation amplitude A plotted against the 'equivalent sample radius' R_s in double logarithmic scale for the 16 subsamples; *b)* the corresponding slopes plotted against the logarithm of R_s (see text).

In Fig. 6a the correlation lengths r_0 obtained from the amplitudes of $1+\xi$ for each subsample are plotted against R_s. Here each r_0 was obtained taking the γ-root of the corresponding amplitude with $\bar{\gamma} = 0.7$. In Fig. 6b we plot the same quantities where now each r_0 is evaluated using its corresponding γ_i ($i = 1, 16$). In each case a power law relation is detectable and a linear fit gives in the first case a slope $m = 1.5 \pm 0.1$ (m here denotes the slope of the power law scaling relation of the correlation length r_0 with the sample size R_s) and in the second case $m = 0.4 \pm 0.1$. This discrepancy is due to the observed trend of the correlation slopes with the sample depth. Clearly in order to compare the results with Eq. (12), which implies that the slope must be unity, its is necessary to get a reliable estimate of γ, and better statistics than the simple least–square method must be adopted. Let us to observe explicitly that the latter result coincides with Davis et al.'s (1988), who report a relation $r_0 \propto \sqrt{R_s}$.

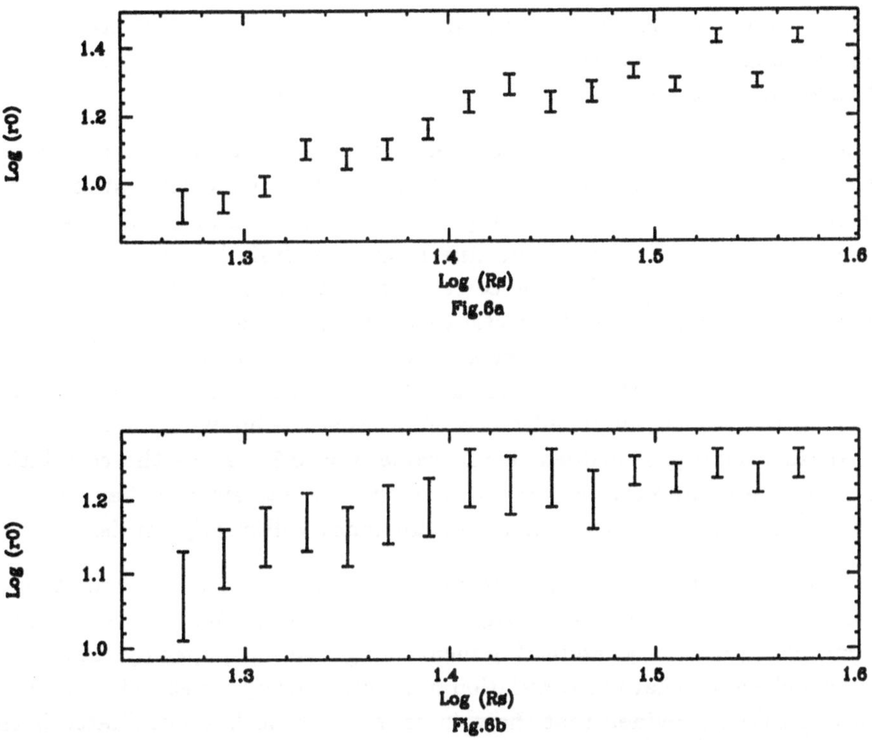

Fig. 6. *a)* the logarithm of the 16 spatial correlation length r_0's plotted as function of $Log(R_s)$. Here each r_0 is obtained as the γ-root of the corresponding amplitude, assuming $\gamma = 0.7$; *b)* the same as *a)*, but now each r_0 is evaluated from its amplitude A by using ithe corresponding value of γ (see text).

It is interesting to observe that also for the small scale regime ($r < r_t$) the correlation function may still be modelled by a power law, also if the approximation is here worse than before. We will report details of this elsewhere, but it is interesting to observe that this is what Guzzo et al. (1991) have found in the Giovanelli-Haynes redshift survey.

6. Conclusion and Summary

The analyses have shown: 1) the use of $1 + w(\theta)$ and of $1 + \xi$ as indicators of the clustering properties seem to be more appropriate than $w(\theta)$ or $\xi(r)$ to describe the galaxy distribution; 2) these 'revisited' correlation functions and their scaling laws with the sample depth reveal the existence of self-similarity in structures which extend on scales ranging from some Mpc up to something like $30 - 40h^{-1}$ Mpc.

The correlation dimension is $D_{2_{sp}} \approx 2.3$ if the results from spatial correlations are considered, while it decreases to $D_{2_{an}} \approx 1.7$ when considering the angular analysis. Actually the spatial analysis gives informations on galaxy clustering properties in the redshift space and not in the real space, so that the effects of the peculiar velocity field are not taken into account. In any case the results here are that the clustering in galaxy distribution seems to be less strong, but more extended in space than what was accepted on the basis of previous results.

As told in §2, an argument against a self-similar distribution has been raised by Davis et al. (1988) about the square root dependence of the correlation length with the sample depth. In fact from Eq. (12) we can see that for a self-similar distribution this scaling law must be linear. Our analysis has shown that the discrepancy is due to the trend of the correlation slope with R_s (Figs. 5b and 6b) which is not expected in a simple fractal case. We have seen that this is mainly due to the relation existing between the amplitude and the slope of the correlation function, which reflects on the fit of the data. This has the effect of increasing the index of the scaling of the amplitudes and of lowering the index of the correlation lengths. As an example, if instead of the value $\bar{\gamma} = 0.7$ we use the correlation slope of the deepest samples, i.e. $\gamma \approx 0.8$, then the scaling law of r_0 has an index $m = 1.3 \pm 0.1$. A fair estimate of the power law index will clarify this issue.

As we have seen from Eq. (12), within the fractal region the distribution of galaxies is characterized by self-similarity and the correlation function is scale free. In this connection, it is very interesting mentioning a recent paper by Ramella et al. (1990), which shows that the correlation function of groups of galaxies coincides with that of galaxies provided that the volume in which both are evaluated is the same. From Eq. (12) it is clear indeed that the spatial correlation function in a self-similar distribution is independent of the hierarchical order of clustering (i.e. if galaxies, groups or clusters), and it depends only on γ and R_s (Calzetti et al. 1988).

As far as the transition to the large-scale homogeneity is concerned, a cut-off value $(32 \pm 6)h^{-1}$ Mpc has been found from the angular analysis (Calzetti et al. 1990); the spatial correlation data do not permit to draw any definite conclusion besides that in the intermediate regime the self-similar scaling seems to continue to the deepest survey ($R_s = 35$ Mpc) and the correlation functions sensibly flatten for a separation of the order of 40 Mpc in each of the 16 samples. This suggests that the cut-off must be searched in a range of the order of $30 - 40h^{-1}$ Mpc, a conclusion similar to that reached by Guzzo et al. (1991) analyzing the Giovanelli-Haynes survey. We have no information on the duration of the transition region. This means that the region within which self-similarity holds is at least $70h^{-1}$ Mpc in diameter.

References

Abell, G.O. (1958): *Astrophys. J. Suppl.* **3**, 211.
Abell, G.O. (1962): *Problems of Extragalactic Research*, Ed. C.G. McVittie (Macmillan, New York), p. 232.
Abell G.O., Corwin Jr., H.G., Olowin, R.P. (1989): *Astrophys. J. Suppl.* **70**, 1.
Atmanspacher, H., Scheingraber, H., Wiedenmann, G. (1989): *Phys. Rev. A* **40**, 3954.
Babul, A., Postman, M. (1990): *Astrophys. J.* **359**, 280.
Bahcall, N.A. (1988): *Ann. Rev. Astron. Astrophys.* **26**, 631.
Bahcall, N.A., Batuski, D.J., Olowin, R.P. (1988): *Astrophys. J. Letters* **333**, L13.
Bahcall, N.A., Soneira, R.M. (1982a): *Astrophys. J. Letters* **258**, L17.
Bahcall, N.A., Soneira, R.M. (1982b): *Astrophys. J.* **262**, 419.
Bahcall, N.A., Soneira, R.M. (1983): *Astrophys. J.* **270**, 20.
Batuski, D.J., Bahcall, N.A., Olowin, R.P., Burns, J.R. (1989) *Astrophys. J.* **341**, 599.
Bean, A.J., Efstathiou, G., Ellis, R.S., Peterson, B.A., Shanks, T. (1983): *Monthly Not. Roy. Astron. Soc.* **205**, 605.
Binggeli, B., Sandage, A., Tamman, G.A. (1988): *Ann. Rev. Astron. Astrophys.* **26**, 509.
Blanchard, A., Alimi, J.M. (1988): *Astron. Astrophys.* **203**, L1.
Börner, G., Mo, H.J. (1990): *Astron. Astrophys.* **227**, 324.
Calzetti, D., Einasto, J., Giavalisco, M., Ruffini, R., Saar, E. (1987): *Astrophys. Sp. Sci.* **137**, 1.
Calzetti, D., Giavalisco, M., Ruffini, R. (1988): *Astron. Astrophys.* **198**, 1.
Calzetti, D., Giavalisco, M., Ruffini, R. (1989): *Astron. Astrophys.* **226**, 1.
Calzetti, D., Giavalisco, M., Ruffini, R., Taraglio, S., Bahcall, N.A. (1991): *Astron. Astrophys.*, in press.
Chincarini, G. (1978): *Nature* **272**, 515.
Chincarini, G., Rood, H.J. (1975): *Nature* **257**, 294.
Chincarini, G., Rood, H.J. (1976): *Astrophys. J.* **206**, 30.
Chincarini, G., Rood, H.J., Thompson, L.A. (1981): *Astrophys. J. Letters* **249**, L47.
Chincarini, G., Vettolani, G. (1987): in *Observational Cosmology*, Eds. A. Hewitt, G. Burbidge, L.Z. Fang (D. Reidel Publ. Co., Dordrecht), p. 275.
Coleman, P.H., Pietronero, L., Sanders, R.H. (1988): *Astron. Astrophys.* **200**, L32.
Davis, M., Meiksin, A., Strauss, M.A., Nicolaci da Costa, L., Yahil, A. (1988): *Astrophys. J. Letters* **333**, L9.
Davis, M., Peebles, P.J.E. (1983): *Astrophys. J.* **267**, 465.
de Lapparent, V., Geller, M.J., Huchra, J.P. (1986): *Astrophys. J.* **302**, L1.
de Lapparent, V., Geller, M., Huchra, J. (1988): *Astrophys. J.* **332**, 44.
de Lapparent, V., Geller, M.J., Huchra, J.P. (1989): *Astrophys. J.* **343**, 1.
Einasto, J., Joeveer, M., Saar, E. (1980): *Monthly Not. Roy. Astron. Soc.* **193**, 353.
Einasto, J., Klypin, A.A., Saar, E. (1986): *Monthly Not. Roy. Astron. Soc.* **219**, 457.
Frisch, U., Parisi, G. (1985): in *Turbulence and Predictability of Geophysical Flows and Climatic Dynamics*, Eds. N. Ghil, R. Benzi, G. Parisi (North Holland, Amsterdam), p. 84.
Gamow, G. (1948): *Nature* **162**, 680.
Groth, E.J., Peebles, P.J.E. (1977): *Astrophys. J.* **217**, 385.
Guzzo, G., Iovino, A., Chincarini, G., Giovanelli, R., Haynes, M.P. (1991): *Nature*, in press.
Halsey, T.C., Jensen, M.H., Kadanoff, L.P., Procaccia, I., Shraiman, B.I. (1986): *Phys. Rev. A* **33**, 1141.

Haynes, M.P., Giovanelli, R. (1986): *Astrophys. J. Letters* **306**, L55.
Huchra, J.P. (1987): private communication.
Huchra, J.P., Davis, M., Lathan, D., Tonry, J. (1983): *Astrophys. J. Suppl.* **52**, 89.
Kirshner, R.P., Oemler, A., Schecter, P.L., Schectmann, S.A. (1981): *Astrophys. J. Letters* **248**, L57.
Klypin, A.A, Kopylov, A.I. (1983): *Sov. Astr. Letters* **9**, 41.
Lahav, O., Rowan-Robinson, M., Lynden-Bell, D. (1988): *Monthly Not. Roy. Astron. Soc.* **234**, 677.
Lilje, P. B. (1988): PhD. Thesis, Cambridge Univ.
Limber, D.N. (1953): *Astrophys. J.* **117**, 134.
Lonsdale, C.J., Helou, G., Good, J.C., Rice, W.L. (1989): *Catalogue of Galaxies and Quasars Observed in the IRAS Survey, Version 2* (Jet Propulsion Lab., Pasadena).
Maddox, S.J., Efstathiou, G., Sutherland, W.J., Loveday, J. (1990): *Monthly Not. Roy. Astron. Soc.* **242**, P43.
Mandelbrot, B.B. (1974): *J. Fluid. Mech.* **62**, 331.
Mandelbrot, B.B. (1975): *C. R. Acad. Sc. Paris* **CCLXXX A**, 1551.
Mandelbrot, B.B. (1979): *C. R. Acad. Sc. Paris* **CCLXXXVIII A**, 81.
Mandelbrot, B.B. (1983): *The Fractal Geometry of Nature* (Freeman, San Francisco).
Martinez, V.J., Jones, B.T (1990): *Monthly Not. Roy. Astron. Soc.* **242**, 517.
Oort, J.H. (1983): *Ann. Rev. Astron. Astrophys.* **21**, 373.
Peebles P.J.E. (1980): *The Large-Scale Structure of the Universe* (Princeton Univ. Press, Princeton).
Peebles, P.J.E., Hauser, M.G. (1974): *Astrophys. J. Suppl.* **28**, 19.
Peebles, P.J.E., Groth, E.J. (1975): *Astrophys. J.* **196**, 1.
Pietronero L. (1987): *Physica* **144A**, 257.
Ramella, M., Geller, M.J., Huchra, J.P. (1990): *Astrophys. J.* **353**, 51.
Ruffini, R., Song, D.J., Taraglio S. (1988): *Astron. Astrophys.* **190**, 1.
Schecter, P.L. (1976): *Astrophys. J.* **203**, 297.
Shane, C.D., Wirtanen, C.A. (1967): *Publ. Lick Obs.* **22**, 1.
Tarenghi, M., Chincarini, G., Rood, J.H., Tompson, L.A. (1980): *Astrophys. J.* **235**, 724.
White, S.D.M (1979): *Monthly Not. Roy. Astron. Soc.* **186**, 145.
Wiedenmann, G., Atmanspacher, H. (1990): *Astron. Astrophys.* **229**, 283.
Zwicky, F., Herzog, E., Wild, P., Karpowicz, M., Kowal, C.T. (1961–68): *Catalogue of Galaxies and of Clusters of Galaxies* (California Inst. Technol., Pasadena).

Fractal Aspects of Galaxy Clustering

Vicent J. Martínez

Departament de Matemàtica Aplicada i Astronomia,
Universitat de València, E–46100 Burjassot, Spain

1. Introduction

In the past decade, the mathematical concept of fractal has exerted a great influence in a large variety of scientific disciplines. It is very common to find recent papers on the application of fractals to different fields in Physics, Chemistry, Biology, etc. The success of the fractal geometry in the description of many systems is due to the fact that deep insights into very simple objects show how fractal measures are more natural for their study.

Mandelbrot's books (1977, 1982) have presented lots of open questions to the scientific community, because they have been written by the author 'without attempting completeness'. This fact, together with the interdisciplinary character of the books has pushed many scientists to embark on the exciting fractal world. Other important readings on the application of fractals in different fields are the books by Barnsley (1988), Feder (1988) and Takayasu (1989).

Astronomy, and in particular, Cosmology has been one of the first fields where fractals have been applied, (Mandelbrot 1975, 1977, 1982; Soneira and Peebles 1978; Efstathiou, Fall and Hogan 1979; Peebles 1980). This fact is easily understood on the grounds of the definition of a fractal set. Although Mandelbrot's statement says that 'a fractal is a set with topological dimension strictly smaller than its Hausdorff dimension', more general and qualitative definitions are used in the application of fractals: 'a fractal set looks similar to itself at every scale', or 'the parts resemble the whole', or 'the density of the object decreases as a power law of the radius'. Many more objects that one thinks at first sight appear to fit one or the other of these intuitive definitions.

The fractal approach is mathematically rigorous and applicable in very different domains. The largest system studied by means of fractal tools is the distribution of galaxies. Several motivations lie behind this fact:

1. The view of the projected Universe. When at the end of the sixties, astronomers had at their disposal large catalogues of galaxies (Zwicky et al. 1961-1968), they could see a map like the one projected in Fig. 1. The image

presents a clumpy distribution of galaxies which could inspire some people to postulate a power-law decreasing behaviour for the density in concentric spheres as a function of the radius (de Vaucouleurs 1970). This is the characteristic behaviour expected in hierarchical fractals.

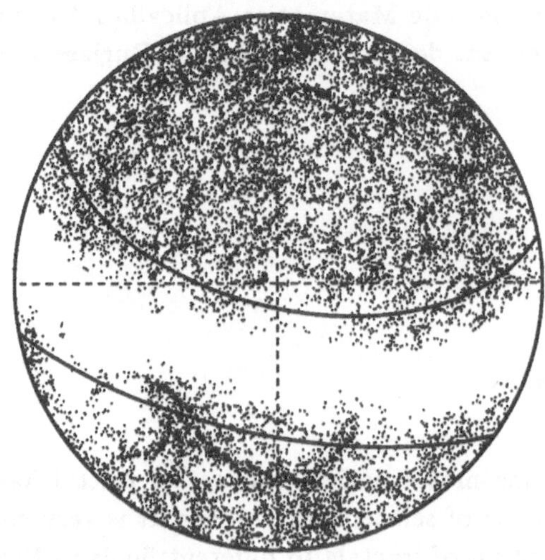

Fig. 1. Equal-area projection of 29363 galaxies from the Zwicky catalogue. The outer circle corresponds to equatorial declination $\delta = 0°$. A solid line marks a band for low Galactic latitudes $|b^{II}| < 20°$, where the catalogue is more empty because of the obscuration due to the interstellar dust of the Milky Way.

2. In the framework of the galaxy formation models, a scale-free power spectrum is generally assumed. The power spectrum is $\langle|\delta_{\boldsymbol{k}}|^2\rangle$ where $\delta_{\boldsymbol{k}}$ is the Fourier transform of the density fluctuations $\delta(\boldsymbol{r}) = [\rho(\boldsymbol{r}) - \langle\rho\rangle]/\langle\rho\rangle$. There is no physical motivation to think about preferred scales in the galaxy formation process, therefore a power-law behaviour for the power spectrum

$$\langle|\delta_{\boldsymbol{k}}|^2\rangle \propto k^n \tag{1}$$

is commonly accepted. For a value of the spectral index $n = 1$, we have the so-called Harrison-Zel'dovich spectrum (Zel'dovich 1972). This is a fractal spectrum in the sense that the corresponding mass fluctuations $\delta M/M$ are independent of the mass-scale when they come through the horizon[1], i.e., every perturbation has at that epoch the same amplitude. The proposed value for the mass variance is about $\sim 10^{-4}$, because in the adiabatic scenario such a value gives rise to fluctuations of the cosmic Microwave Background Radiation of the order of 10^{-4}.

[1] When the perturbations come through the horizon, the Zel'dovich spectrum becomes proportional to k^{-3}.

3. Clustering of galaxies in redshift space. In the past recent years, several catalogues of galaxies, listing the redshift of each object, have been completed (Tully and Fisher 1978; Huchra et al. 1983; Giovanelli et al. 1986). The presence of large filaments and sheets, together with big voids almost devoid of luminous matter has been clearly established (Kirshner et al. 1981, 1987; Bhavsar and Ling 1988; Geller and Huchra 1989). The plot of the famous 'slice of the Universe' (see Fig. 2) (de Lapparent, Geller and Huchra 1986) is very similar to some fractal patterns formed in aggregation processes of solid particles such as the so-called cluster-cluster aggregation (Pietronero and Tosatti 1986, and references therein). But, obviously, the underlying physical processes which are responsible for the clustering in each case are completely different. However, in both cases, fractal measures can help in the characterization of the clustering properties. In Cosmology, the study of the large-scale patterns observed in the galaxy distribution, as it can be appreciated in Fig. 2, plays an important role in recent fractal approaches (Saar and Saar 1990).

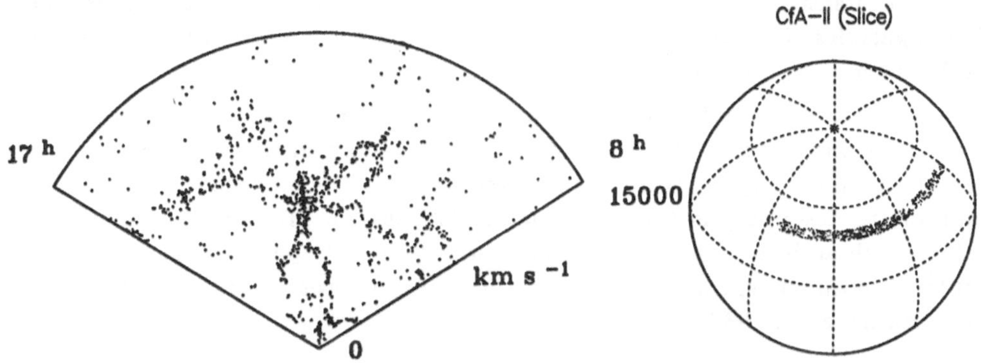

Fig. 2. The 'slice of the Universe' (de Lapparent et al. 1986) and the part of the celestial sky surveyed in equatorial coordinates: $8^h \leq \alpha \leq 17^h$ and $26.5° \leq \delta \leq 32.5°$. The catalogue is complete down to apparent magnitude $m_B = 15.5$ and contains 1057 galaxies.

4. The power-law correlation function. Finally, I have to mention a quantitative reason for thinking in a possible fractal galaxy distribution. The spatial two-point correlation function $\xi(r)$ is defined in terms of the probability of finding a neighbour of a randomly selected galaxy, in a volume d^3r at distance r of that galaxy,

$$dP = \bar{n}(1 + \xi(r))d^3r \tag{2}$$

$\bar{n}$ being the average number density of galaxies. The correlation function (Davis and Peebles 1983) has been found to obey a characteristic power-law behaviour

$$\xi(r) = \left(\frac{r}{r_0}\right)^{-\gamma} \tag{3}$$

for $r < 10h^{-1}$ Mpc (h being the Hubble constant in units of 100 km s^{-1} Mpc^{-1}). The parameters of the power-law are $\gamma = 1.77 \pm 0.04$ and $r_0 =$

$5.4 \pm 0.3\,h^{-1}$ Mpc. This result was also deduced from the analysis of the angular correlation function $\omega(\theta)$ of the projected data (Groth and Peebles 1977). One can calculate $\xi(r)$ from $\omega(\theta)$ by means of the Limber equation (Peebles 1980). A galaxy distribution with a power-law correlation function leads to the idea of a fractal, more precisely one can conclude that the galaxy distribution has a characteristic fractal dimension[2] $D \simeq 3 - \gamma = 1.23 \pm 0.04$.

In view of the previous arguments, one could think that a fractal description of the Universe is free from drawbacks. However, it should be noted that in the standard cosmology, the distribution of mass must tend to a non-zero finite density when averaged over large volumes. There are not only theoretical arguments in favour of the large–scale homogeneity, there is also a strong observational evidence from the extremely homogeneous and isotropic measurements of the Microwave Background Radiation. A fractal Universe without crossover to homogeneity (Mandelbrot 1989) implies a vanishing density for very large volumes and this idea cannot be accepted without creating important additional problems. Thus, an essential point is the extent of the fractal regime in the spatial distribution of galaxies. This question and the evidence of a breakdown in the scaling will be discussed in the following sections.

The second main point of this work is the character of the fractal measure for the galaxy distribution. We will show how the galaxy distribution, even in the fractal or scaling region, is not well represented by a homogeneous fractal. The appropriate descriptor is the multifractal distribution. Multifractals are fractal objects with an invariant measure characterized not only by a relevant dimension but by a continuum spectrum of singularities, (Mandelbrot 1974; Frisch and Parisi, 1985; Halsey et al. 1986).

2. Fractal Models for the Cosmic Structure

Since the pioneering work of Fournier d'Albe (1907) and Charlier (1922) on hierarchical models for the Universe, other fractal models have been proposed to approximate the distribution of galaxies in the space (Mandelbrot 1975; Soneira and Peebles 1978). In this section, we are going to show how these more recent fractal textures look when projected onto the celestial sphere.

[2] To be rigorous, this value corresponds to the correlation dimension D_2, which could be different of other fractal dimensions (see Sect. 7).

2.1 Rayleigh-Lévy Dust

This model, introduced by Mandelbrot in 1975, is based on a Rayleigh-Lévy flight. It works as follows: let us consider a starting point of the simulation; from here, following a random walk we jump over to a second point. The constraint affects only the distribution of the jump lengths, while the direction of each jump is taken isotropically at random. If X denotes the random variable which contains the values of the jump lengths, its probability distribution function has to be a power-law

$$P(X > \epsilon) = \left(\frac{\epsilon}{\epsilon_0}\right)^{-D} \tag{4}$$

If at each jump we place a galaxy the output looks as in Fig. 3. The density of this distribution of points within a sphere of radius R varies as R^{D-3}. This behaviour provides us with a first definition of fractal dimension. In a fractal object the *mass* (in our case the number of points) scales with the radius as $M(R) \propto R^D$. Note that for a homogeneous distribution on a line we get $M(R) \propto R^1$, if the distribution is placed uniformly on a surface, the behaviour is $M(R) \propto R^2$, etc. In a fractal object the exponent D is strictly smaller than the dimension d of the Euclidean space in which the object is embedded, therefore the density scales as $\rho(R) \propto R^{D-d}$.

This kind of fractal served Mandelbrot to define the concept of *lacunarity* (Mandelbrot 1982). The lacunarity is related to the presence of large empty regions in the fractal. This fact agrees qualitatively with the observational evidence of the existence of very large voids almost devoid of galaxies (Kirshner et al. 1981, 1987), but the size of the empty regions in the model are much larger than in the observations. Thus, the lacunarity of the model is too high to be accepted.

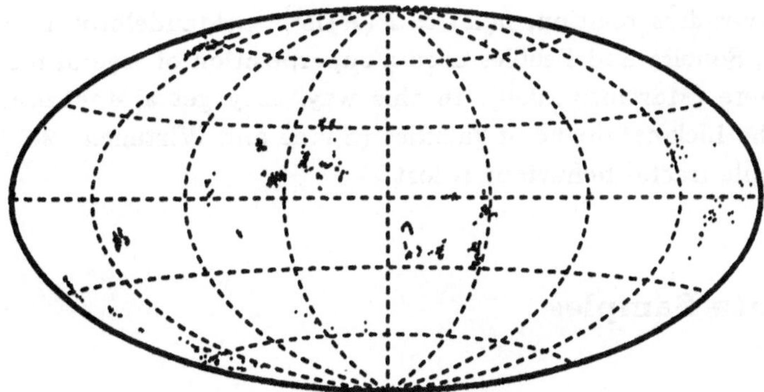

Fig. 3. Hammer's equal-area projection of the Rayleigh-Lévy fractal dust. The fractal dimension D (see Eq. (4)) has been chosen to be $D = 1.23$.

2.2 Soneira-Peebles Hierarchical Model

This model provides a distribution of points in the three-dimensional space defining a simple fractal. If the parameters are appropriately selected, the correlation function of the model reproduces the observed power-law (Eq. (3)). We build the model as follows: place randomly in a sphere of radius R the centers of η spheres of radius R/λ with $\lambda > 1$. Now, in each one of these spheres we place again randomly the centers of η new spheres of radius R/λ^2 and so on. If this recursive process is repeated indefinitely we get a homogeneous fractal. In any stage of the construction, each sphere is similar to whichever of the others and looks like the whole set under the appropriate magnification. If an object of size R is formed by $\eta = \lambda^{D_S}$ similar objects of size R/λ, the similarity dimension is D_S, therefore the similarity dimension of this distribution depends on the parameters η and λ in the following way

$$D_S = \frac{\log \eta}{\log \lambda} \tag{5}$$

If we truncate this construction at level L and place a galaxy in the centers of the spheres belonging to the last generation, the resulting point set is a fractal set (see Fig. 4). It has been shown that the density of this fractal within a sphere of radius R follows a power-law $\rho(R) \propto R^{-\gamma}$ (Peebles 1980) with $\gamma = 3 - \log \eta / \log \gamma$. Therefore the density-radius dimension defined above is equal, in this case, to the similarity dimension. In order to illustrate how this model looks like when projected on the celestial sphere, we have chosen the parameters $\eta = 2$ and $\lambda = 1.76$ which give rise to a dimension $D = 1.23$. The number of levels is $L = 13$, then 8192 points have been placed. In Fig. 4 an equal-area projection of this fractal has been plotted. Again, the distribution presents a high degree of lacunarity. A mathematical aspect of the lacunarity is related to the possible existence of a non constant prefactor in the density-radius relation, $\rho(R) \propto P(R)R^{D-d}$ (Mandelbrot 1989). To avoid this problem, Soneira and Peebles used a superposition of several models like the one shown here (Martínez 1990). In this way, they get a stochastic simulation resembling the Lick catalogue of galaxies (Shane and Wirtanen 1967). But doing that, the simple fractal behaviour is lost.

3. The Data Samples

To see how fractals and multifractals are applied to the study of the large-scale structure of the Universe, we are going to deal with real data. Two available redshift surveys from the Harvard-Smithsonian Center for Astrophysics (CfA) have been used for this analysis.

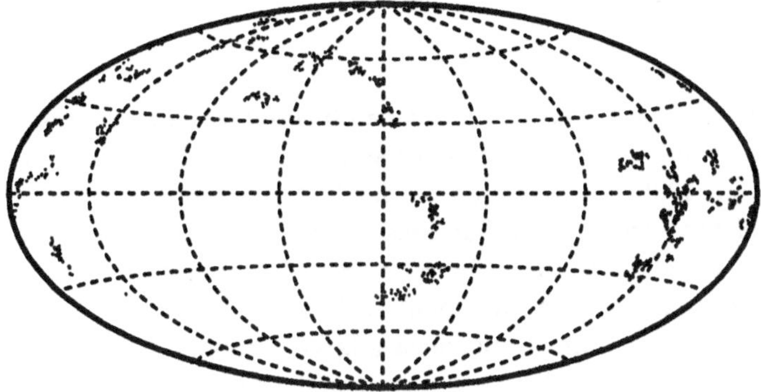

Fig. 4. Hammer's equal-area projection of the Soneira and Peebles hierarchical model. The parameters have been fixed to get a simple fractal point distribution with the same dimension as the one plotted in Fig. 3, $D = 1.23$.

3.1 CfA-I

The compilation of redshifts referenced as CfA-I (Huchra et al. 1983) covers two different areas of the sky. The first region lies on the northern hemisphere with Galactic latitude $b^{II} \geq 40°$ and declination $\delta \geq 0°$ and the second region lies on the southern sky with $b^{II} \leq -30°$ and $\delta \geq -2.5°$. The surveyed solid angles are 1.83 and 0.84 steradians respectively. The catalogue is complete down to the apparent magnitude $m_B = 14.5$. We will work only at the northern part which contains nearly 2000 galaxies. Due to the fact that the catalogue has been built with a fixed apparent magnitude limit m_B, galaxies with exactly that magnitude would not be in the survey if their distance to us was larger. Hence, the catalogue is not uniform from the point of view of the luminosity. Using the redshift as a distance indicator, we have decided to extract complete volume-limited subsamples. This is done by keeping only galaxies brighter than an absolute magnitude

$$M_B^{\max} = m_B - 25 - 5\log(V_{\max}/H_0), \tag{6}$$

for a sample with maximum depth $D_{\max} = V_{\max}/H_0$. The resulting sample is now uniform regarding luminosity, but an enormous amount of collected data is not considered. For example, a subsample with an absolute magnitude limit $M_B^{\max} = -19.5 + 5\log h$ which corresponds to a maximum depth $D_{\max} \simeq 63.1\,h^{-1}$ Mpc contains only 412 galaxies or a subsample with $100\,h^{-1}$ Mpc depth and formed only by galaxies brighter than $-20.5+5\log h$ contains 220 objects. We will use these two subsamples in the following sections. They will be called $S63$ and $S100$ respectively. We have used a version of the CfA catalogue with the redshifts corrected as in Einasto et al. (1984). These corrections are due to solar motion, Virgocentric flow, peculiar velocities and a small relativistic effect.

3.2 CfA-II (Slice)

The plot shown in Fig. 2 was first published in 1986 by de Lapparent et al., but the data has been available for the use of the astronomical community after publication in Huchra et al. 1990. This version of the CfA redshift survey is complete down to $m_B = 15.5$ but covers a smaller area of the celestial sphere (see Fig. 2). The sample lies on the strip bounded by $8^h \leq \alpha \leq 17^h$ in right ascension and $26.5° \leq \delta \leq 32.5°$ in declination. The increment with respect to the CfA-I in one unity of the apparent magnitude limit represents a large increase of the number of galaxies in the surveyed region. In Fig. 5 we have plotted the absolute magnitude of all the galaxies of the sample (1057) against their radial velocities. The lines show the absolute magnitude limit as a function of the distance corresponding to the apparent magnitude m_B (see Eq. (6)). The solid line corresponds to $m_B = 15.5$ while the dashed line corresponds to $m_B = 14.5$. The number of galaxies below the dashed line is only 184, which means that changing from 14.5 to 15.5 in the apparent magnitude limit, the number of galaxies seen in the area increases nearly six times.

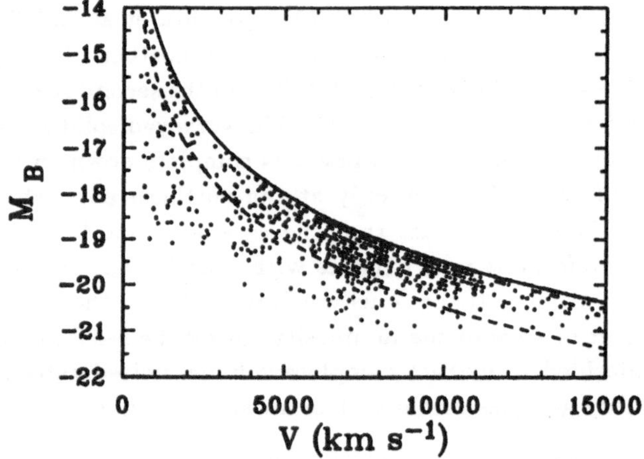

Fig. 5. Absolute magnitude versus radial velocity for the sample CfA-II (Slice).

Different authors have performed the statistical analysis of these samples by using a luminosity function to overcome the problem of missing fainter galaxies with the depth, (de Lapparent et al. 1988, 1990). This kind of analysis was also done for the CfA-I survey by Davis and Peebles (1983), with the luminosity function derived in Davis and Huchra (1982). The advantage of this method is that one can use the whole data set in some statistical analysis. Usually it is assumed that there exists a universal function expressing the number of galaxies per unit volume in the luminosity interval $[L, L + dL]$. This quantity is denoted $\Phi(L)dL$, where the luminosity function $\Phi(L)$ is generally parametrized to fit the Schechter (1976) functional

$$\Phi(L)dL = \Phi_* \left(\frac{L}{L_*}\right)^{\alpha} \exp\left(-\frac{L}{L_*}\right) d\left(\frac{L}{L_*}\right). \tag{7}$$

In this scheme the number density $\bar{n}$ and the estimation of $\xi(r)$ by means of counting pairs must be done as a function of the selection criteria, which depends on the luminosity function. Usually corrections in the counts are made by giving each galaxy a weight inversely proportional to the probability of being included in the magnitude-limited sample.

Other authors prefer to use complete volume-limited samples for the statistical studies (Einasto et al. 1984, 1986; Klypin et al. 1989; Maurogordato and Lachièze-Rey 1987, 1990; Domínguez-Tenreiro and Martínez 1989). We are going to do the same choice here, because our aim is to present more transparently the fractal techniques in cosmology. In the next section, we will show how to estimate $\xi(r)$ in this scheme. From the CfA-II (Slice) we select a subsample formed only by galaxies with radial velocity $V \leq 10000$ km s^{-1} and absolute magnitude $M_B \leq -19.5 + 5\log h$. The subsample contains 205 galaxies. It is plotted in the standard way in Fig. 6. We call it $SL100$.

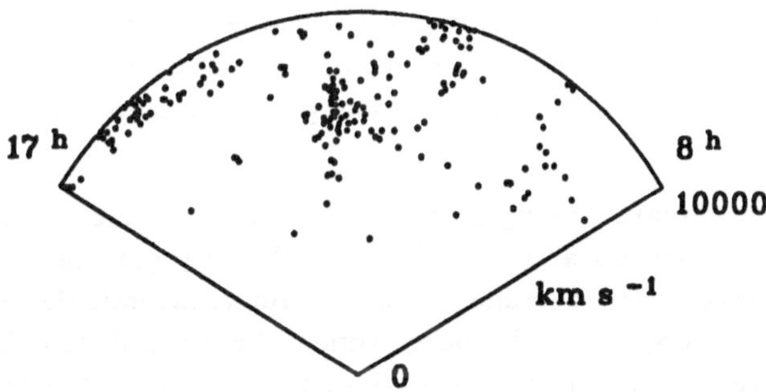

Fig. 6. A complete volume-limited subsample drawn from the CfA-II (Slice).

4. Correlation Dimension

The standard approach to the study of the large-scale statistical properties of the galaxy clustering has been the use of the two-point correlation function, (Peebles 1980). Higher order N-point correlation functions can be defined as extensions of Eq. (2), but it is practically impossible to measure these functions for $N \geq 4$ (Bonometto and Sharp 1980).

Grassberger and Procaccia (1983) defined the correlation dimension as the exponent of a power-law correlation integral

$$C(r) = \int_0^r 4\pi s^2[1 + \xi(s)]ds \propto r^{D_2}. \tag{8}$$

Then, if $1+\xi(r)$ behaves as a power-law r^{D_2-3} we can easily get an estimation of the correlation dimension. In fact, as we have already mentioned $\xi(r) \propto r^{-\gamma}$, then for $r \ll r_0$, where $\xi(r) \gg 1$, the correlation dimension is $D_2 \simeq 3-\gamma$. However, in order to estimate D_2, it is much better to calculate directly the slope of the plot $\log(1+\xi(r))$ vs. $\log(r)$ in the scaling region, (Coleman et al. 1988; Martínez 1990). In any case, we need to find a practical way to measure $\xi(r)$ for a finite point set.

4.1 Practical Calculation of $\xi(r)$

The method presented here in the calculation of the two-point correlation function is based on the fact, that for an infinite volume of space, the correlation function is given by the average

$$1+\xi(r) = \frac{1}{N}\sum_{i=1}^{N} \frac{N_i(r)}{\bar{n}4\pi r^2 dr}, \tag{9}$$

where $N_i(r)$ is the number of points in a spherical shell of thickness dr and radius r centered at the point i and $\bar{n}$ is the mean number density. The average is taken over a large number of points N. In our case, we deal with finite volumes, the number density is just $\bar{n} = N/V$, where V denotes the whole volume and N is the total number of points in the sample. Now, the boundary effects must be carefully considered in the calculation. There are two possibilities to circumvent the problem of the edges.

1. If the sample is large enough, one can consider an inner region surrounded by a buffer zone (Upton and Fingleton 1987). Only galaxies of the inner region are considered in the average of Eq. (9) but those galaxies are allowed to *see* neighbours belonging to the outer zone. This procedure is illustrated in Fig. 7a. The great problem of this method is the fact that it is wasting a lot of relevant information since an important proportion of the point set may lie in the buffer zone. If the total number of points is relatively small, the actual number of points involved in the average (9) will be too low to be acceptable.
2. The second method is to replace the volume of the spherical shell in Eq. (9) $4\pi r^2 dr$ by the part of that volume which lies within the sample space (Rivolo 1986). Let us call this volume $V_i(r)$. When the geometry of the sample volume is complicated, Monte Carlo algorithms are generally used to calculate $V_i(r)$. Eq. (9) becomes now

$$1+\xi(r) = \frac{V}{N^2}\sum_{i=1}^{N} \frac{N_i(r)}{V_i(r)}. \tag{10}$$

Fig. 7b illustrates this method. In this technique, one is implicitly assuming that the galaxy distribution is an isotropic and stationary random point process, (Peebles 1989).

Obviously the first method can only be applied to models of the distribution of galaxies which provide enough number of points. For real catalogues, where the number of data is rather small, the second procedure is more appropriate.

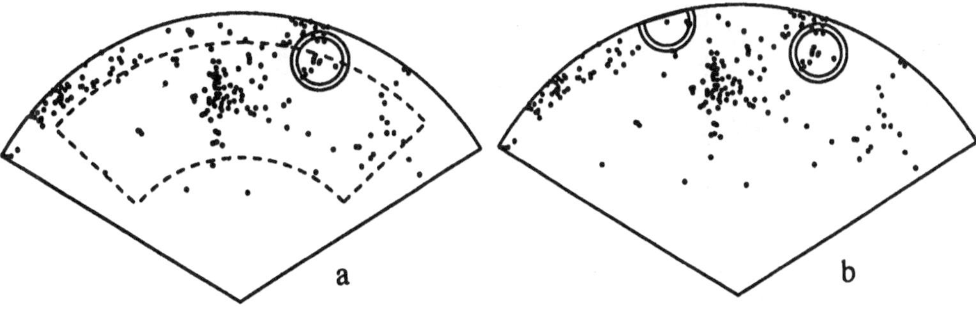

Fig. 7. Two methods for border corrections: a) considering an outer buffer zone; b) calculating intersections of shell volumes with the sample space.

4.2 Correlation Analysis of the CfA Samples

When we apply the previous technique to the calculation of the two-point correlation function of the samples[3] presented in Sect. 3, we get the result shown in Fig. 8. As we see, there exists a scaling region where a power-law with exponent ~ -1.7 is a good representation of $1+\xi(r)$. Therefore, we conclude that the correlation dimension is about ~ 1.3 for the galaxy distribution. The discrepancy in the amplitudes of the three curves is due to the fact that the sample $S100$ is formed by galaxies intrinsically brighter than the galaxies belonging to the other two samples. Galaxies brighter than $M_c \simeq -20 + 5\log h$ tend to be more clustered than fainter galaxies (Davis et al. 1988; Hamilton 1988; Domínguez-Tenreiro and Martínez 1989). This effect is usually called luminosity segregation.

Although the samples $S63$ and $SL100$ are formed by galaxies with the same luminosity limit, the subslice is dominated by extremely important density enhancements, as the Coma cluster or part of the Great Wall (Geller and Huchra 1989). Obviously, this kind of rich features are not so common if we consider larger volumes. The influence of the local density peaks in the amplitude of the correlation function is not negligible (Martínez et al. 1991). We can asses that the correlation length of CfA samples is affected by both luminosity effects and presence of local inhomogeneities. However, there are other interpretations of the increment of the amplitude of $\xi(r)$ with the depth of the sample (see lines corresponding to $S63$ and $S100$). Pietronero (1987) and Coleman et al. (1988) suggested that the possible linear increase of r_0 with sample depth is the proof of an unbounded simple fractal distribution of galaxies, at least up to the present observational limits.

[3] We have removed from the complete subsamples of CfA-I, those galaxies with radial velocities less than $V_{\min} = 1700$ km s^{-1}.

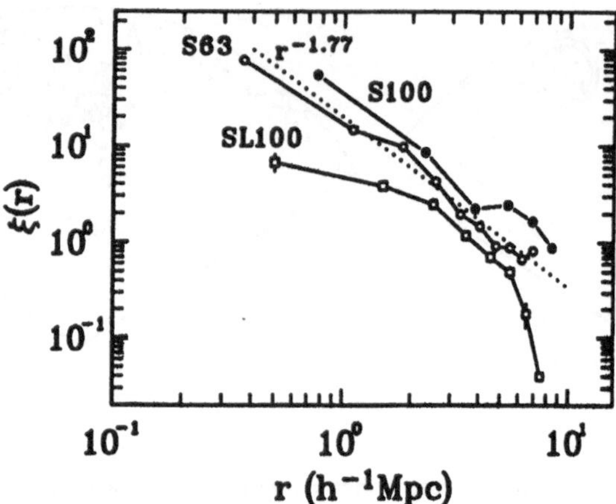

Fig. 8. The two-point correlation function for the three CfA subsamples (solid lines). The dotted line shows the standard fit $\xi(r) = (r/5.4)^{-1.77}$ (Davis and Peebles 1983).

5. Fractals and the Variation of Density

In Sect. 2, we have already mentioned the density-radius relation expected in a fractal object $\rho(R) \propto R^{D-3}$. Usually, the calculation of D by means of this formula is done by taking the average when the positions of the centers of the spheres change. This averaging is qualitatively equivalent to the formula used in the calculation of the correlation function (see Eq. (9)). However, due to the finiteness of the sample and the boundary effects, one cannot go beyond a certain distance in this type of calculations. For example, for $r > 20h^{-1}$ Mpc the results will not be reliable anymore if we take into account the size of the present samples. Nevertheless, one can go to further distances without doing the average, by considering how the density varies in concentric volumes centered at our Galaxy. The result of this procedure is obviously biased by the choice of the center, but, for very large volumes, the effects of local inhomogeneities will probably be diluted. Of course, now we can study the density-radius relation for larger distances.

5.1 Transition to Homogeneity

The most interesting question is whether the density will be continuously decreasing when the volume increases, as expected for an unbounded fractal, or on the contrary, a roughly constant mean density exists when averaged over large volumes. To answer this question, we have performed the calculation with the two deeper samples $S100$ and $SL100$. In Fig. 9 we show the results. For the large volume $S100$, one can see that beyond the overdensity corresponding to the Virgo cluster ($R < 20\,h^{-1}$ Mpc), there exists a scaling zone where the power-law holds (Martínez and Jones 1990), but, for larger distances, a clear tendency to a constant density is also appreciated on the figure. However, for the subslice the density does not follow any power-law. The effects of the local inhomogeneities of this sample can be easily seen in the density profile (density in concentric shells) which is marked with dashed lines. The Coma cluster and the Great Wall are clearly noticed at 70 and 95 h^{-1} Mpc respectively. The average density is $3.6 \times 10^{-4}\, h^3$ Mpc^{-3} for the sample $S100$ and $5.1 \times 10^{-3}\, h^3$ Mpc^{-3} for $SL100$.

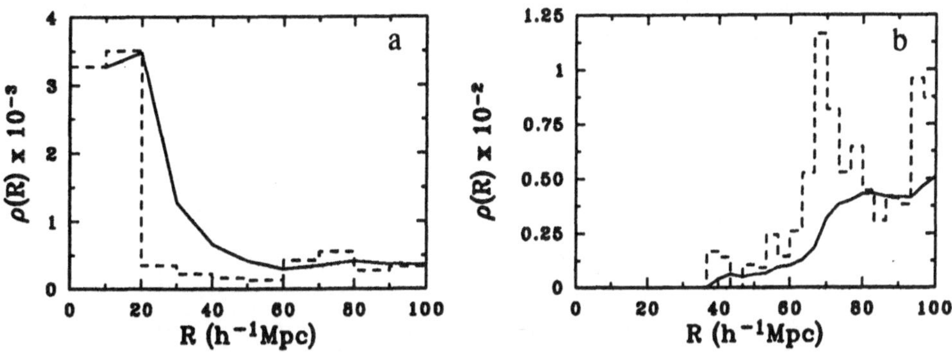

Fig. 9. The density-radius relation (solid lines) for the sample $S100$ (a), and for the sample $SL100$ (b). The dashed lines show the corresponding density profile.

Taking together the results shown here and in Sect. 4, it seems reasonable to suggest a picture for the large-scale structure of the Universe characterized by a transition between a fractal regime at small scales to homogeneity at larger distances. Several models have been presented to simulate the galaxy clustering reproducing precisely this transition (Calzetti et al. 1988; Castagnoli and Provenzale 1991). This kind of behaviour is also observed in the majority of dynamical models of galaxy formation. As an illustration, we deal here with one of the most promising models which presents this transition. The model is based on a new version of the so-called *pancake* scenario (Buchert 1989a).

5.2 The Pancake Model

This model is based on a new class of solutions of the Euler-Poisson equations for describing the formation of structure (Buchert 1989b). The new theory is a generalization of the 'Zel'dovich approximation' (Zel'dovich 1970, 1978) that was the only theory to approximate the growing of the density perturbations in the non-linear regime ($\delta\rho/\rho \gg 1$) by means of an extrapolation of the linear theory of gravitational instability (Lifshitz 1946). From high resolution simulations of Buchert's model, we can extract luminous galaxies by a procedure called dynamical thresholding. The idea is qualitatively similar to the biasing galaxy formation models (Bardeen et al. 1986) which have a clear application in cosmological N-body simulations in cold dark matter scenarios (White et al. 1987). In the pancake model, the suggested procedure is slightly different: following the trajectories of the model particles (elementary fluid volumes) in their Lagrangian evolution, we identify those volumes as *luminous* if their absolute density exceeds a particular density threshold χ_c. If this condition is fulfilled in some moment of their Lagrangian history, we mark the point and keep it as a galaxy to be plotted in the Eulerian space at $z = 0$. This has been done in Fig. 10, where we show a 2D sample containing 26028 particles after thresholding with $\chi_c = 10$. The model was originally formed by 512^2 particles (as tracers of the baryonoic matter), and flat power-law spectrum with cut-off length $\lambda_{\min} = 25\,h^{-1}$ Mpc. The sample size corresponds to 48 times $\lambda_{\min}$ (Buchert 1991).

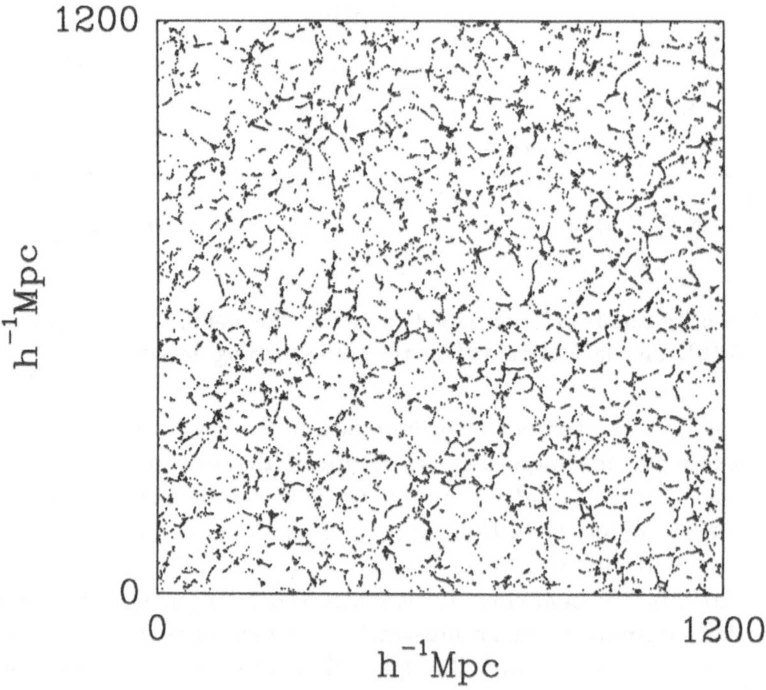

Fig. 10. A 2D simulation of a high resolution pancake model after dynamical thresholding with $\chi_c = 10$.

What is interesting from this model in the present context is the compatibility between a fractal regime at small scales with large-scale homogeneity. In Fig. 11a, we show the pair distribution function $1 + \xi(r)$ for this model, evaluated in the range $[0, 50\,h^{-1}$ Mpc]. The calculations have been performed following the technique explained in Sect. 4.1 with the buffer zone to overcome the boundary effects. While, for small distances ($r < 10\,h^{-1}$ Mpc), the function fits well a power-law (dotted line), it is evident that for larger scales it deviates from this behaviour. In fact, one can see that if r is large enough $1+\xi(r) \rightarrow 1$, indicating a clear tendency to homogeneity (Buchert and Martínez 1991). The function relating the density in concentric circles with its radius is plotted in Fig. 11b. The circles are centered in the closest point to the center of the square. The peaks of the curve correspond to the intersections of the circles with the pancakes, this behaviour is related with the quasi-periodicity observed in deep pencil beam redshift samples of galaxies (Broadhurst et al. 1990; Buchert and Mo 1990). However, the curve is smoothed if we take the average of the variation of density when the centers of the circles change (dotted line in Fig. 11b).

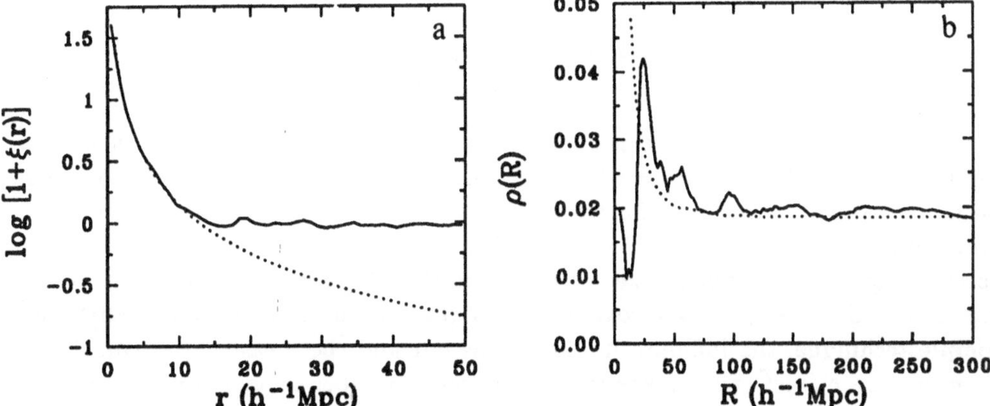

Fig. 11a. The correlation function for the pancake model shown in Fig. 10 (solid line). The dotted line shows fractal behaviour.

Fig. 11b. The density-radius relation for the same model (solid line). The dotted line is the same function after averaging over many centers.

6. Hausdorff Measures

At the beginning of this work, we have mentioned Mandelbrot's first definition of a fractal in terms of the Hausdorff dimension. However, this mathematically well defined concept introduced by Hausdorff at the beginning of the century (Hausdorff 1917) is not usually taken into account in the applications of fractals. The difficulties in the direct use of the Hausdorff dimension will become clear after its definition.

6.1 Hausdorff Dimension and Capacity

Let us consider a set A embedded in the Euclidean space $\Re^n$. The family of all countable coverings of A formed by sets with diameter less or equal than a given positive number $r > 0$ is

$$\Upsilon_A^r = \{\{B_i\}_{i\in I} \mid A \subseteq \bigcup_{i\in I} B_i \mid r_i \leq r\} \tag{11}$$

B_i being the sets belonging to each r-covering of A, with diameter r_i. The β-dimensional outer measure of A is defined as

$$H^{\beta}(A) = \lim_{r\to 0} \inf_{\Upsilon_A^r} \sum_i r_i^{\beta}. \tag{12}$$

In sum, we consider the moments of the diameters of the covering sets, take the infimum among all possible coverings and find the limit when the upper bound of the diameters vanishes. The Hausdorff dimension is defined as a critical exponent for which the measure $H^{\beta}(A)$ changes from zero to infinity,

$$H^{\beta}(A) = \begin{cases} \infty & \text{if } \beta < D_H(A) \\ 0 & \text{if } \beta > D_H(A) \end{cases} \tag{13}$$

The rigorous mathematical treatment of this concept can be found in the book by Falconer (1985). Note that the definition (12) involves looking at all coverings of the set and selecting that for which the moments take the minimal value. This procedure is too complicated to be applied to arbitrary sets. The standard solution is covering the space with boxes of the same size. Nevertheless the result of this method will be only an approximation to the Hausdorff dimension. With this simple technique, the infimum in (12) drops out and the measure has the simple form $\lim_{r\to 0} N(r)r^{\beta}$, where $N(r)$ is the minimum number of boxes of size r needed to cover A. As we expect this measure to be a non-zero finite number (see Eq. (13)), the best solution is to find a number $D_C(A)$ such that

$$N(r) \sim \left(\frac{1}{r}\right)^{D_C(A)}. \tag{14}$$

If this exponent exists, $D_C(A)$ is the capacity of the set, which can be more rigorously defined (Kolmogorov and Tihomirov 1959) by

$$D_C(A) = \lim_{r\to 0} \frac{\log N(r)}{\log(1/r)}. \tag{15}$$

Nevertheless, definition (14) is more suitable for practical applications of fractals, because in any physical system there is always a smallest characteristic scale, which renders the limit involved in (15) meaningless. In practice, the capacity or box-counting dimension is estimated as the slope of the plot $\log(N(r))$ vs. $\log(1/r)$. In any case, it can be shown that inequality $D_H(A) \leq D_C(A)$ holds for any set, but the equality is true in the majority of fractals.

If we want to apply these measures to the description of the cosmic structure, an additional problem must also be considered. Our set is a finite point set. Then, its Hausdorff dimension is trivially zero and the capacity, even with the operational definition given in (14), may be dramatically affected by poor statistics (Jones et al. 1988). In the following paragraphs we give some answers.

6.2 The Minimal Spanning Tree

Although we have at our disposal a point set, we may consider such set as a point process lying on an uncountable support. We want to estimate the dimension of the support using only the data from the finite sample. This may be done by means of the Minimal Spanning Tree (MST), (Martínez and Jones 1990; van de Weygaert et al. 1990). The MST of a set of N points is the unique graph connecting all the points without closed loops and with minimum length. In the present context, we will consider this construction as the 'minimal' covering of the set. This can be used as an approximation to the infimum in (12). Then, we have a way for a direct estimation of the Hausdorff dimension: we select randomly n points from the total sample and calculate the MST of this subset, now we consider the lengths of the $m = n - 1$ branches of the tree $\{\ell_i\}_{i=1}^{m}$, as the diameters of the covering. Then, if the moments of these lengths behave as

$$\sum_{i=1}^{m} \ell_i^{\beta}(m) \sim m^{1-\beta/h(\beta)} \tag{16}$$

(when the number of random points n changes), the fixed point of the function $h(\beta)$ estimates the Hausdorff dimension, $h(\tilde{D}_H) = \tilde{D}_H$ (see Eq. (13)).

We have applied this technique to the sample $SL100$. In Fig. 12 we show how the MST connects all the points of the sample.

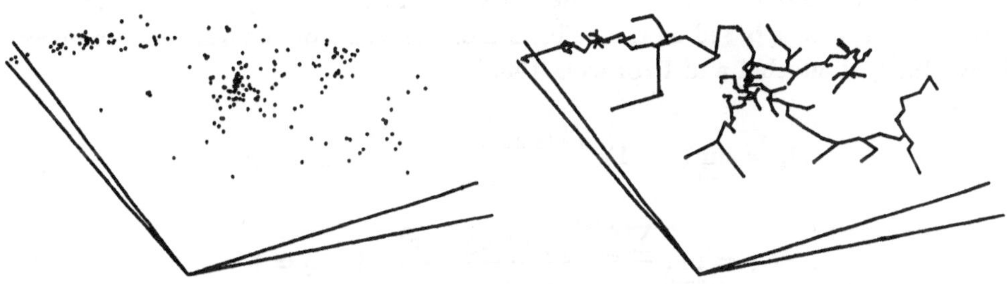

Fig. 12. The Minimal Spanning Tree of the sample $SL100$.

The calculation yields a value of $\tilde{D}_H = 2.0 \pm 0.1$ in agreement with previous calculations performed with the survey CfA-I (Martínez and Jones 1990), and also in agreement with other estimations of the dimensionality by means of different methods (Saar 1989, Martínez 1990, de Lapparent et al. 1990). The value of the dimensionality ~ 2 reveals the dominant role of the sheet-like structures.

The difference between the correlation dimension $D_2 \simeq 1.3$ and the Hausdorff dimension $D_H \simeq 2.0$ may be quite well explained in the context of multifractals. In next section we are going to present a more general theory of dimensions which is relevant when the invariant measure is not uniformly distributed on the fractal.

7. Multifractals

Simple fractal objects are rather special in being characterized by only one dimension. A broad set of different phenomena verifying some kind of scaling properties need more than one single number to be fully characterized (Stanley and Meakin 1988). The generalizations of fractals which describe this kind of anomalous scaling are called multifractals (Paladin and Vulpiani 1987). From the different scaling laws associated to the moments of a fractal measure, we will be able to extract an infinite hierarchy of exponents. Each one of these exponents is related with the fractal dimension of a subset of the whole fractal object. Hence, multifractals can be interpreted as an inextricable mixture of simple fractals, each one characterized by its own fractal dimension and the whole set described by an infinite family of relevant exponents.

Multifractals were first introduced by Mandelbrot in the analysis of turbulence (Mandelbrot 1974). More modern applications of this formalism deal with the characterization of the strange attractors which appear in some non-linear dynamical systems (Halsey et al. 1986). Jones et al. (1988) introduced the use of multifractal techniques in the analysis of the large-scale galaxy distribution.

7.1 Generalized Dimensions

Multifractals can be defined as a generalization of the Hausdorff dimension (Halsey et al. 1986, Martínez 1990) or as a generalization of the capacity. This second way is more practical and is based on the definition of the Rényi or generalized dimensions (Rényi 1970; Hentschel and Procaccia 1983),

$$D_q = \lim_{r \to 0} (q-1)^{-1} \frac{\log \sum_{i=1}^{N(r)} \rho_i(r)^q}{\log r} \quad \text{if} \quad q \neq 1 \tag{17}$$

$$D_1 = \lim_{r \to 0} \frac{\sum_{i=1}^{N(r)} \rho_i(r) \log \rho_i(r)}{\log r} \quad \text{if} \quad q = 1 \tag{18}$$

where $\rho_i(r)$ is the probability that a point of the set lies in cell i, if the set has been covered by disjoint cells of equal size r. Obviously, for $q = 0$, D_0 is the capacity defined in (15).

For a homogeneous fractal $D_q = D_0$ for all q, instead in a multifractal set, D_q is a monotonic decreasing function of q. The numbers $D_{+\infty}$ and $D_{-\infty}$ are the generalized dimensions corresponding to the densest parts of the set and to the more rarefied parts respectively. This can be easily seen because if $q \to +\infty$, the partition sum in (17) is dominated by the largest $\rho_i(r)$ value; likewise, if $q \to -\infty$, the sum is dominated by the lowest non-zero $\rho_i(r)$.

Another standard representation of the multifractal characteristics is the so-called $f(\alpha)$ spectrum. The relation between D_q and $f(\alpha)$ is given by a Legendre transform.[4]

[4] The pair (α, f) is the Legendre transform of the pair $(q, (q-1)D_q)$.

Multifractals provide us with an extremely powerful tool to distinguish point set distributions. In the following example we show how this ability for discriminating fractal patterns works.

7.2 Random Multiplicative Multifractal Lattice

This model was proposed to describe turbulent processes (Meakin 1987). It has the advantage of having an analytically calculable D_q function. The construction of the model works as follows: we divide a square into four equal square pieces, assigning to each one a probability p_1, p_2, p_3 and p_4. Next, we divide again each of the new squares into four subsquares and assign to each one a probability which is the product of one of the p_i's taken at random (without repetition) by the p_i corresponding to the parent square. We continue this procedure for a number of levels L. Thus, we get a $2^L \times 2^L$ lattice and each pixel has attached a value of the type $\mu = p_1^i p_2^j p_3^k p_4^l$ with $i + j + k + l = L$. Then, we distribute a large number of points on the lattice with probabilities for each pixel just proportional to those numbers. In Fig. 13 we show two realizations of this process with the choice of the parameters,

I. $p_1 = 1$, $p_2 = 1$, $p_3 = 1$, $p_4 = 0$
II. $p_1 = 1$, $p_2 = 0.67$, $p_3 = 0.446$, $p_4 = 0.116$

Although the two fractal patterns look completely different to the eye, both have the same correlation dimension D_2. In fact, the first example is a simple fractal, while the second has a multifractal distribution. The Rényi dimensions of this model depend only on the choice of the initial parameters $\{p_i\}_{i=1}^4$ in the following way

$$D_q = (1-q)^{-1} \log_2(f_1^q + f_2^q + f_3^q + f_4^q) \tag{19}$$

where $f_i = p_i / \sum_{i=1}^4 p_i$, (if $f_j = 0$, the term f_j^q is removed from (19)). Model I yields $D_q = D_0 = \log 3/\log 2$ for all q and therefore it is a simple fractal. Instead, model II presents a non-trivial D_q function. Both curves are also shown in Fig. 13, below the corresponding pattern. The multifractal approach quantifies the differences between the two pictures, while the correlation dimension alone, being the same for both cases, is not able to distinguish the two sets.

7.3 Algorithms for the Estimation of the Generalized Dimensions

Box-counting methods are not too efficient to get the generalized dimensions. Alternative procedures have already been proposed (Grassberger et al. 1988; Martínez et al. 1990). We briefly summarize these techniques.

1. In the so-called correlation method, we center a sphere of radius r on each galaxy of the sample ($i = 1, \ldots, N$) and count the number of galaxies within this ball, $n_i(r)$. The probability at scale r associated with the galaxy i is $p_i(r) = n_i(r)/N$, then the expected multifractal scaling law is

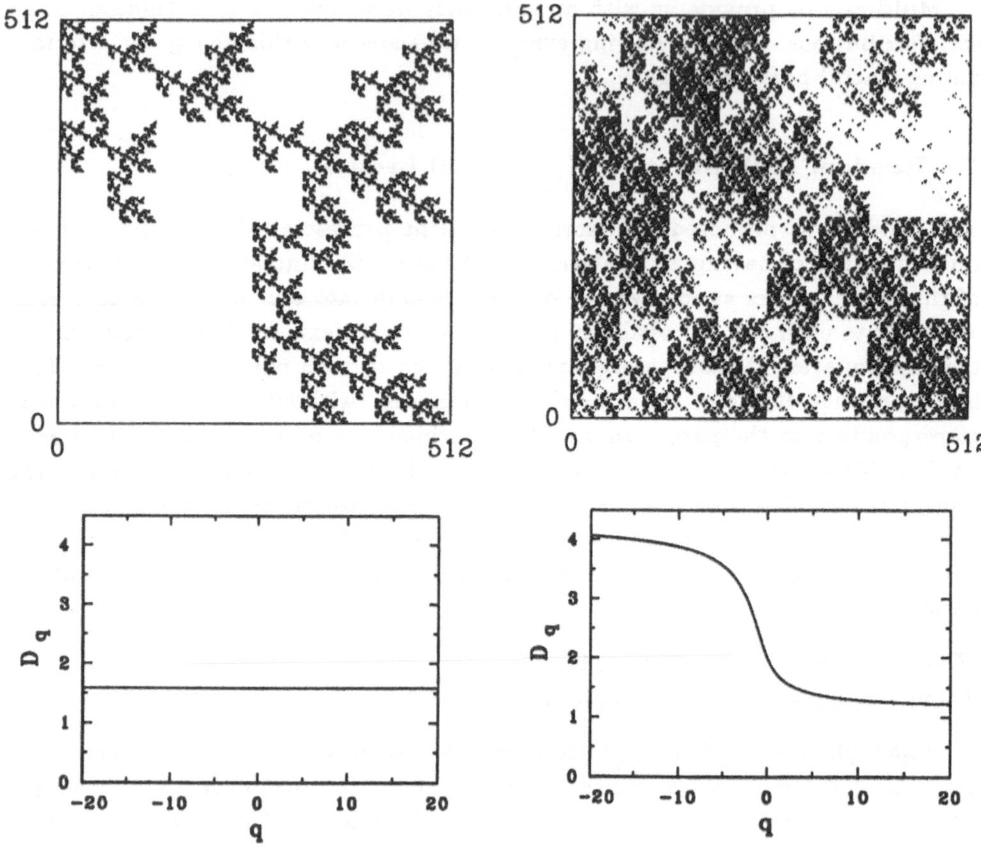

Fig. 13. On the upper panels we show two realizations of the multiplicative random cascade. The left panel is a simple fractal, while the right is a multifractal. Below each panel, the corresponding D_q function has been plotted.

$$\frac{1}{N}\sum_{i=1}^{N} p_i(r)^{q-1} \sim r^{(q-1)D_q} \times \text{const.} \tag{20}$$

2. Another practical method is based on balls with fixed mass. This is done by surrounding each galaxy with the smallest sphere that encompasses a given number of points, (or a given probability p). Let us call the radius of this sphere $r_i(p)$. The generalized dimensions are now calculated by fitting the expression

$$\frac{1}{N}\sum_{i=1}^{N} r_i(p)^{-(q-1)D_q} \sim p^{1-q} \times \text{const.} \tag{21}$$

3. A new technique based on the Minimal Spanning Tree is now available. It comes from a generalization of (16) in the following way,

$$\frac{1}{m}\sum_{i=1}^{m} \ell_i(m)^{-(q-1)D_q} \sim m^{q-1} \times \text{const} \tag{22}$$

where $m = n - 1$, and n was the number of points in a subsample randomly selected from the whole sample. This number varies in a range $1 < N_1 \leq n \leq N_2 < N$, where the power-law holds.

The efficiency of the first two methods has been successfully tested in galaxy samples (Domínguez-Tenreiro and Martínez 1989; Martínez et al. 1990). The first one, based on spheres with fixed radius, works better for $q > 1$, while the second one, based on spheres with fixed mass, gives better results for $q < 1$.

The new formalism, based on the MST, has now given the first positive results in the analysis of the multifractal properties of strange attractors (van de Weygaert et al. 1990, Domínguez-Tenreiro et al. 1990). Unfortunately, this method needs a large number of points N to be applied for $|q| \gg 0$. We have calculated the function D_q of the galaxy sample $S63$ by means of the first two algorithms. The result is shown in Fig. 14. As we can see, the galaxy distribution in the scaling region does not obey a simple fractal law, but it follows a multifractal distribution, (Jones et al. 1988; Atmanspacher et al. 1989; Martínez et al. 1990).

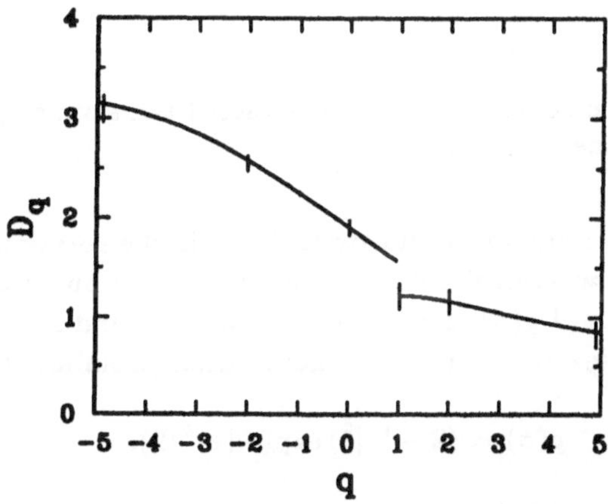

Fig. 14. Generalized dimensions of the sample $S63$. The multifractal character is evident because D_q is not constant. Different algorithms have been used for $q < 1$ and for $q > 1$.

7.4 Wavelets

An alternative approach to multifractals by means of the wavelet transform has been recently suggested (Arneodo et al. 1988; Argoul et al. 1989). The wavelet integral transform of a function $\mu(x)$ with respect to the wavelet g is

$$T_g(a, b) = \frac{1}{a^n}\int g^*(a^{-1}r^{-1}(x - b))\mu(x)dx \tag{23}$$

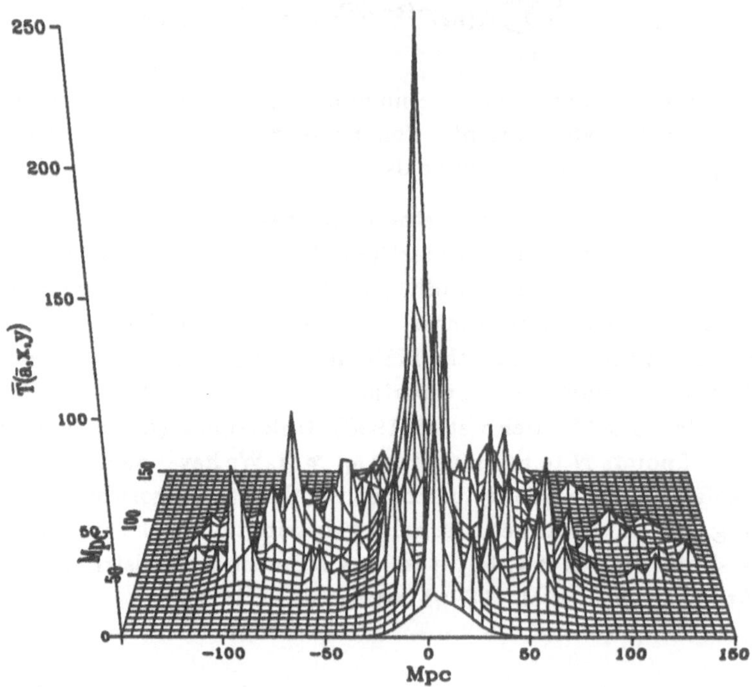

Fig. 15. The superposition of the maxima of the wavelet transforms $T_g(a, b)$ for different a values, calculated on the CfA-II slice.

where * stands for the complex conjugate and r^{-1} is the rotation operator. Some of the moments of the wavelet function g must vanish, in any case, it has to be a function with zero mean ($\int g(\boldsymbol{x})d\boldsymbol{x} = 0$) and localized around zero. As the standard wavelet, the radical 'Mexican Hat' is commonly used (Combes et al. 1988)

$$g(\boldsymbol{x}) = (2 - |\boldsymbol{x}|^2)\exp(-|\boldsymbol{x}|^2/2). \tag{24}$$

Wavelets act as 'mathematical microscopes' (Arneodo et al. 1988). The magnification of the structure at position $\boldsymbol{b}$ is given by the factor a^{-1}. Wavelets have also been used in Astronomy as identifiers of structure (Slezak et al. 1990) and in the analysis of star-forming regions (Gill and Henriksen 1990).

In the way that wavelets work, different structures of the data sample appear to be more relevant if a^{-1} changes. For example, increasing the magnification a^{-1}, the small-scale structures are enhanced. In order to visualize simultaneously the structures at all scales, we can do a superposition of the wavelet transforms for different values of a and consider at each position $\boldsymbol{b}$ the maximum of $T_g(a, \boldsymbol{b})$. To illustrate this formalism, we have applied it to the CfA-slice shown in Fig. 2, (Martínez and Saar 1991). As a first approximation, we have considered all the data on a 2D plane ($\boldsymbol{b} = (x, y)$). The plot of the surface showing the superposition of wavelet transforms corresponding to different values of a is shown in Fig. 15.

The rich structures of the slice are nicely visualized by means of this approach. The multifractal characteristics may be extracted from the expected scaling laws $|T_g(a,b)| \sim a^{\alpha(a)}$ (Arneodo et al. 1988; Martínez and Saar 1991).

Acknowledgements

I thank my colleagues Thomas Buchert, Rosa Domínguez-Tenreiro, Bernard Jones, Enn Saar and Rien van de Weygaert for the enjoyable working days lived together. Most of the ideas appearing here come from the fruitful collaboration with them. I thank the Institut de Física Corpuscular (València) for computing facilities. This work was partially supported by the Dirección General de Investigación Científica y Técnica (project number PB86-0292-C04-04), Spain.

References

Argoul, F., Arneodo, A., Elezgaray, J., Grasseau, G., Murenzi, R. (1989): *Phys. Lett. A* **135**, 327.

Arneodo, A., Grasseau, G., Holschneider, M. (1988): *Phys. Rev. Lett.* **61**, 2281.

Atmanspacher, H., Scheingraber, H., Wiedenmann, G. (1989): *Phys. Rev. A* **40i**, 3954.

Bardeen, J.M., Bond, J.R., Kaiser, N., Szalay, A.S. (1986): *Astrophys. J.* **304**, 15.

Barnsley M. (1988): *Fractals Everywhere*, (Academic Press, San Diego).

Bhavsar, S.P., Ling, N. (1988): *Astrophys. J. Letters* **331**, L63.

Bonometto, S.A., and Sharp, N.A. (1980): *Astron. Astrophys.* **92**, 222.

Broadhurst, T.J., Ellis, R.S., Koo, D.C., Szalay, A.S. (1990): *Nature* **343**, 726.

Buchert, T. (1989a): *Rev. Mod. Astr.* **2**, 267.

Buchert, T. (1989b): *Astron. Astrophys.* **223**, 9.

Buchert, T., Mo, H. (1990): *Astron. Astrophys.*, in press.

Buchert, T. (1991): in *Physical Cosmology*, (Frontières, Paris), in press.

Buchert, T., Martínez, V.J. (1991): in preparation.

Calzetti, D., Giavalisco, M., Ruffini, R. (1988): *Astron. Astrophys.* **198**, 1.

Castagnoli, C., Provenzale, A. (1991): *Astron. Astrophys.*, in press.

Charlier, C.V.L. (1922): *Ark. Math. Astr. Fys.* **16**, 1.

Coleman P.H., Pietronero, L., Sanders R.H. (1988): *Astron. Astrophys.* **200**, L32.

Combes, J.M., Grossman, A., Tchamitchian, P. (Eds.) (1988): *Wavelets* (Springer–Verlag, Berlin).

Davis, M., Huchra, J. (1982): *Astron. J.* **254**, 437.

Davis, M., Peebles, P.J.E. (1983): *Astrophys. J.* **267** 465.

Davis, M., Meiskin, A., Strauss, M. A., da Costa, L. N., Yahil, A. (1988): *Astrophys. J.* **333**, L9.

de Lapparent, V., Geller, M.J., Huchra, J.P. (1986): *Astrophys. J. Letters* **302**, L1.

de Lapparent, V., Geller, M.J., Huchra, J.P. (1988): *Astrophys. J.* **332**, 44.

de Lapparent, V., Geller, M.J., Huchra, J.P. (1990): preprint.

de Vaucouleurs, G. (1970): *Science* **167**, 1203.

Domínguez-Tenreiro, R., Martínez, V.J. (1989): *Astrophys. J. Letters* **339**, L9.

Domínguez-Tenreiro, R., Roy, L. J., Martínez, V.J. (1990): preprint.

Efstathiou, G., Fall, S.M., Hogan, G. (1979): *Monthly Not. Roy. Astron. Soc.* **189**, 203.

Einasto, J., Klypin, A.A., Saar, E., Shandarin, S.F. (1984): *Monthly Not. Roy. Astron. Soc.* **206**, 529.

Einasto, J., Klypin, A.A., Saar, E. (1986): *Monthly Not. Roy. Astron. Soc.* **219**, 457.

Falconer, K.J. (1985): *The Geometry of Fractal Sets*, (Cambridge Univ. Préss, Cambridge).

Feder, J. (1988): *Fractals*, (Plenum Press, New York).

Fournier d'Albe, E. E. (1907): *Two New Worlds*, (Longmans Green, London).

Frisch,U., Parisi, G. (1985): in *Turbulence and Predictability of Geophysical Flows and Climatic Dynamics*, Eds. N. Ghil and R. Benzi, (North Holland, Amsterdam).

Geller, M.J., Huchra J.P. (1989): *Science* **246**, 897.

Gill, A.G., Henriksen, R.N. (1990): preprint.

Giovanelli, R., Haynes, M.P., Chincarini, G.L. (1986): *Astrophys. J.* **300**, 77.

Grassberger, P., Procaccia, I. (1983): *Phys. Rev. Lett.* **50**, 346.

Grassberger P., Badii R., Politi, A. (1988): *J. Stat. Phys.* **51**, 135.

Groth, E.J., Peebles, P.J.E. (1977): *Astrophys. J.* **217**, 385.

Halsey, T.C., Jensen, M.H., Kadanoff, L.P., Procaccia, I., Shraiman, B.I. (1986): *Phys. Rev. A.* **33**, 1141.

Hamilton, A.J.S (1988): *Astrophys. J. Letters* **331**, L59.

Hausdorff, F. (1919): *Math. Ann.* **79**, 157.

Hentschel, H.G.E., Procaccia, I. (1983): *Physica* **8D**, 435.

Huchra, J.P., Davis, M., Latham, D., Tonry, J. (1983): *Astrophys. J. Suppl.* **52**, 89.

Huchra, J., Geller, M.J., de Lapparent, V, Corwin, H. (1990): *Astrophys. J. Suppl.* **72**, 433.

Jones, B.J.T., Martínez, V.J., Saar, E., Einasto, J. (1988): *Astrophys. J. Letters* **332**, L1.

Kirshner, R.P., Oemler, A., Schechter, P.L., Shectman, S.A. (1981): *Astrophys. J. Letters* **248**, L57.

Kirshner, R.P., Oemler, A., Schechter, P.L., Shectman, S.A. (1987): *Astrophys. J.* **314**, 493.

Klypin, A.A., Einasto, J., Einasto, M., Saar, E. (1989): *Monthly Not. Roy. Astron. Soc.* **237**, 929.

Kolmogorov, A. N., Tihomirov, V. M. (1959): *Upekhi Matematicheskikh Navk* **14**, 3.

Lifshitz, E. M. (1946): *J. Phys.* **5**, 84.

Mandelbrot, B.B. (1974): *J. Fluid. Mech.* **62**, 331.

Mandelbrot, B.B. (1975): *C.R. Acad. Sc. Paris A* **105**, 1551.

Mandelbrot, B.B. (1977): *Fractals: Form, Chance and Dimension* (Freeman, San Francisco).

Mandelbrot, B.B. (1982): *The Fractal Geometry of Nature* (Freeman, San Francisco).

Mandelbrot, B.B. (1989): in *Large Scale-Structure and Motions in the Universe*, Eds. M. Mezzeti, G. Giuricin, F. Mardirossian, M. Ramella, (Kluwer Acad. Publ., Dordrecht), p. 259.

Martínez, V.J. (1990): in *Fractals in Astronomy*, Ed. A. Heck, *Vistas Astron.* **33**, 337

Martínez, V.J., Jones, B.J.T. (1990): *Monthly Not. Roy. Astron. Soc.* **242**, 517.

Martínez, V.J., Jones, B.J.T., Domínguez-Tenreiro, R., van de Weygaert, R. (1990): *Astrophys. J.* **357**, 50.

Martínez, V.J., Portilla, M., Jones, B.J.T. (1991): in preparation.

Martínez, V.J., Saar, E. (1991): in preparation.

Maurogordato, S., Lachièze-Rey, M. (1987): *Astrophys. J.* **320**, 13.

Maurogordato, S., Lachièze-Rey, M. (1990): preprint.

Meakin, P. (1987): *Phys. Rev. A* **36**, 2833.

Paladin, G., Vulpiani, A. (1987): *Phys. Repts.* **156**, 147.
Peebles, P.J.E. (1980): *The Large-Scale Structure of the Universe* (Princeton Univ. Press, Princeton).
Peebles, P.J.E. (1989): in *Cosmology and Particle Physics*, Eds. J.F Nieves, D.R. Altschuler (World Scientific, Singapore).
Pietronero, L., Tossati, E. (Eds.), 1986, *Fractals in Physics*. (North Holland, Amsteredam).
Pietronero, L. (1987): *Physica A* **144**, 257.
Rényi, A. (1970): *Probability Theory* (North Holland, Amsterdam).
Rivolo, A.R. (1986): *Astrophys. J.* **301**, 70.
Saar, E. (1989): *Proc. 11th Krakow School of Cosmology*, Eds. H. Durbeck, P. Flin (Springer-Verlag, Berlin).
Saar, E., Saar, V. (1990): preprint.
Schechter, P. (1976): *Astrophys. J.* **203**, 297.
Shane, C.D., Wirtanen, C.A. (1967): *Publ. Lick Obs.* **22**, part 1.
Slezak, E., Bijaoui, A., Maars, G. (1990): *Astron. Astrophys.* **227**, 301.
Soneira, R., Peebles, P.J.E. (1978): *Astron. J.* **83**, 845.
Stanley, H.E., Meakin, P. (1988): *Nature* **335**, 405.
Takayasu, H. (1989): *Fractals in the Physical Sciences* (Manchester Univ. Press, Manchester).
Tully R.B., Fisher J.R. (1978): in *The Large-Scale Structure of the Universe*, Eds. M.S. Longair, J. Einasto (Reidel Publ. Co., Dordrecht), p 214.
Upton G.J.G., Fingleton, B. (1987): *Spatial Data Analysis by Example*, Vol. **1** (Wiley, New York).
Van de Weygaert, R., Jones, B.J.T., Martínez, V. J. (1990): preprint.
White, S.D.M., Davis, M., Efstathiou, G., Frenk, C.S. (1987): *Nature* **330**, 451.
Zel'dovich, Ya.B. (1970): *Astron. Astrophys.* **5**, 84.
Zel'dovich, Ya.B. (1972): *Monthly Not. Roy. Astron. Soc.* **160**, 1P.
Zel'dovich, Ya.B. (1978): in *The Large-Scale Structure of the Universe*, edd. M.S. Longair, J. Einasto (Reidel Publ. Co., Dordrecht), p. 409.
Zwicky, F., Herzog, E., Wild, P., Karpowicz, M., Kowal, C.T. (1961-1968): *Catalogue of Galaxies and Clusters of Galaxies* (California Inst. Techn., Pasadena), Vol **1-6**.

Fractal Properties in the Simulations of a One-Dimensional Spherically Expanding Universe

J.L. Rouet, E. Jamin, M.R. Feix

PMMS/CNRS, 3A Avenue de la Recherche Scientifique,
F-45071 Orléans Cedex 2, France

Abstract: A model universe, expanding according to the Hubble law, is numerically simulated. The model consists of a thin central part of a spherically symmetric and homogeneous universe. By appropriate rescaling, the dynamics becomes, in the case of critical expansion, analogous to the dynamics of a one-dimensional one-component plasma with the addition of a friction term. The motion is most appropriately solved in phase space. Physical properties of the model are examined. The results of two simulations corresponding to different initial conditions are presented; they show a hierarchical clustering process of the density with clumping of the velocities in each cluster. The configuration space density and the phase space distribution function are found to have prominent fractal structures, with for the former a fractal dimension independent of initial conditions of $.575 \pm .035$. The importance of phase space in the study of the fractal structure of large dynamic systems is stressed.

0. Introduction

Since the observations made by Hubble, the different models of formation of our universe suppose it homogeneous, isotropic and in expansion (Peebles 1980). Nevertheless, observations by Soneira and Peebles (1977) point out the existence of a hierarchical structure (at least up to a certain scale beyond which the universe is supposed homogeneous). Finally, the density measurements show that the mass contained inside a sphere of radius R, increases as R^D, where D is a number smaller than 3. Consequently some authors have developed fractal models which preserve the topological properties of the universe by selecting a proper D (Mandelbrot 1983).

We proceed differently: we start from a simple physical model for the universe (in fact an N body problem). We study, by numerical simulations, its evolution in phase space and find that this evolution leads to structures having fractal properties and dimension.

We present the selected model in Sect. 1; in Sect. 2 we introduce the rescaled time and space variables more appropriate to describe the Hubble expansion and we give their physical interpretation; Sect. 3 describes our algorithm, and gives the data for the different numerical experiments and their results; these results are discussed and analyzed from a fractal point of view in Sect. 4.

1. Model

The selected model (see also Rouet et al. 1990) is a small portion of a homogeneous universe with spherical symmetry, located far from the center and from the boundary, in the limit of a vanishing ratio of thickness to radius. Particles (in fact stars) are located on concentric spheres, which expand with respect to their common center. The expansion goes according to Hubble's law for which the length scaling factor $C(t)$ varies as $t^{2/3}$. Both in the case of the Friedmann equations (general relativity) as in classical Newtonian mechanics, this expansion is at the limit between a periodic evolution of the universe and a freely expanding one for which the radius increases asymptotically with time (Andrillat 1984). Notice that with this $t^{2/3}$ law, the expansion is self similar for a spherically symmetric model. Moreover, in this case, the presence of a boundary at finite distance (beyond which there is no matter) can be ignored. This is true only for this model with the boundary limit and the scaling factor having the same $t^{2/3}$ time evolution. This property has been demonstrated by Munier et al. (1979) for a collisionless self gravitating system and by Bouquet et al. (1985) in the hydrodynamic case.

The assumption that the thickness of the studied slice of the universe is much less than the distance to the horizon allows us to use the Newtonian approximation. Numerical simulations of a spherical model (Bouquet, private communication) with homogeneous initial conditions and the $t^{2/3}$ expansion scale factor indicate that, progressively, the system divides itself into two parts. In the first we find the 'stars' with low energy, collapsing towards the center to become trapped by the self gravitating potential. In the second part the high energy stars decouple from each other and from the rest of the system with a final free expansion. At the end of the simulation all particles have selected the group to which they belong and none will follow the $t^{2/3}$ expansion law. Here we deal with the particles of the central zone which have neither decoupled from the bulk of the system (high energy stars) nor fallen into the central trap (low energy stars) and consequently still follow the Hubble expansion. The existence of such particles for significant times in the above simulation, indicates that the study of their dynamics is an interesting point to be studied.

In the limit where the size of the studied system is much smaller than the radius of the limiting spheres, we can substitute a plane geometry to the spherical one (the particles being now located on moving expanding planes). This change of geometry presents several advantages:

- there is no center of expansion (a hypothetical center of the universe)

- there is no boundary for such universe and we are really dealing with the large scale homogeneous and isotropic system suggested by the observations.

Nevertheless the expansion treated corresponds to the Hubble one, i.e. a spherically symmetric expansion and it is this expansion which will be absorbed in the rescaling of the space-time variables.

With these assumptions the equation of motion of one particle is simply written

$$\frac{d^2x}{dt^2} = E \tag{1}$$

with $dE/dx = -4\pi G\rho$, where ρ is the homogeneous density and G the gravitational constant and x represents the distance from an arbitrary reference plane to our system.

2. Rescaling

The rescaling group method introduces a correspondence between the old independent variables x and t and the new ones $\hat{x}$ and $\hat{t}$ by means of two functions of time $A(t)$ and $C(t)$ (Besnard et al. 1983):

$$x = C(t)\hat{x} \qquad \text{and} \qquad dt = A(t)^2 d\hat{t}. \tag{2}$$

Note that although the topology is planar, three dimensional rescaling is used. The modification of spheres to plane sheets has just constrained the expansion law in two directions.

We apply the transformation to the gravitational field and the density by writing

$$E(x,t) = B(t)\hat{E}(\hat{x},\hat{t}) \qquad \text{and} \qquad \rho(x,t) = D(t)\hat{\rho}(\hat{x},\hat{t}). \tag{3}$$

At this stage no assumptions have been made on the form of the rescaling functions $A(t)$, $B(t)$, $C(t)$, $D(t)$ and we have complete freedom to choose them. Let us then define the functions $B(t)$ and $D(t)$ in order to keep the physical properties identical. First impose mass conservation:

$$D = C^{-3}. \tag{4}$$

Second, impose the gravitational field to keep the same shape, thus insuring an invariant expression of the Poisson law:

$$B = C^{-2}. \tag{5}$$

Defining the new velocity $\hat{v} = d\hat{x}/d\hat{t}$ and adopting the notation $\dot{z} = dz/dt$, $\ddot{z} = d^2z/dt^2$ the transformed equation of motion (1) takes the form

$$\frac{d^2\hat{x}}{d\hat{t}^2} + 2A^2\left(\frac{\dot{C}}{C} - \frac{\dot{A}}{A}\right)\frac{d\hat{x}}{d\hat{t}} + A^4\frac{\ddot{C}}{C}\hat{x} = \frac{A^4}{C^3}\hat{E}. \tag{6}$$

Two new terms appear in Eq. (6): a term of friction (in $d\hat{x}/d\hat{t}$) and a term corresponding to a new attractive or repulsive force (in $\hat{x}$).

If we choose $A = C$, we cancel the friction term and obtain the transformation adopted by Doroshkevich et al. (1980), but then the coefficients of Eq. (6) still depend on time t. In their work, this time dependence is included in the gravitational constant G which consequently varies with time.

On the contrary, if we look for scale functions such that the coefficients of Eq. (6) become time independent, we adopt:

- $C(t) \sim t^{2/3}$ which is the critical expansion factor. Actually, this is no surprise as this rescaling method often gives border solutions, and this is the case for this self similar solution witch is the classical equivalent of the Einstein-de Sitter universe, which we intend to study here. With this choice of $C(t)$, the motion in the rescaled space is reduced to the residual motion, often called peculiar motion, after substraction of the main Hubble velocity.
- $A(t) \sim t^{1/2}$ defines the new time $\hat{t} \sim ln(t)$.

To clarify the meaning of this new time $\hat{t}$, let us show how to obtain on physical grounds these dependencies of $A(t)$ and $C(t)$. We choose $C(t)$ according to the universe expansion; this means that in the transformed variable $\hat{x}$ the universe becomes stationary in the absence of peculiar motion. To conserve the total mass and homogeneity of the universe, the density should read

$$\rho = C^{-3}\rho_o \tag{7}$$

with ρ_o the initial density. The natural unit of time is chosen as the inverse Jeans frequency ω_j defined by

$$\omega_j^2 = 4\pi G\rho. \tag{8}$$

It is clear that the density and thus the natural time unit ω_j^{-1}, vary with time. Consider a new time $\hat{t}$ measured in constant units of the inverse *initial* Jeans frequency

$$\omega_{jo}d\hat{t} = \omega_j(t)dt \tag{9}$$

where $\omega_{jo}^2 = 4\pi G\rho_o$. With (7) and (8) this relation is equivalently written as

$$d\hat{t} = C^{-3/2}dt \tag{10}$$

and imposes the constraint (see Eq. 1)

$$A(t)^2 = C(t)^{3/2}. \tag{11}$$

If we now make the choice of a critical universe with $C \sim t^{2/3}$, we have $A \sim t^{1/2}$, i.e. the same scale factors as those rendering the coefficients of Eq. (6) time independent.

This physically rescaled time, measured in units of ω_{jo}^{-1}, allows, consequently, for the critical universe, a description of the peculiar motion with a particularly simple equation. With the explicit expressions $C(t) = (\alpha\omega_{jo}t)^{2/3}$ and $A(t) =$

$(\alpha\omega_{jo}t)^{1/2}$, where the arbitrary numerical constant α is taken equal to $3/\sqrt{2}$, Eq. (6) becomes

$$\frac{d^2\hat{x}}{d\hat{t}^2} + \frac{\omega_{jo}}{\sqrt{2}}\frac{d\hat{x}}{d\hat{t}} - \omega_{jo}^2\hat{x} = \hat{E}. \tag{12}$$

The rescaling has replaced the expansion by a new dynamical problem, governed by new forces (friction and repulsion) having constant coefficients, in addition to the invariant gravitational force. We have a system of infinite planes of constant superficial density, moving in a box of constant length, interacting with each other and, additionally, with a uniform neutralizing background of repulsive mass density. This model is analogous to the one-dimensional one-component plasma (O.C.P.) (e.g. Feix 1978). Thus our equation, in the absence of the friction force, is formally the same as that of a O.C.P, provided we replace $(-4\pi G)^{-1}$ by the permittivity of vacuum ϵ_o.

One way to treat such a system is to use the Vlasov equation, which describes the time evolution of the fluid in *phase space*. The dispersion relation, obtained in the linear regime for large wavelengths, reads, in the plasma case,

$$\omega^2 = \omega_p^2 + 3k^2V_T^2 \tag{13}$$

and accordingly, in the gravitational case,

$$\omega^2 = -\omega_{jo}^2 + 3k^2V_T^2 \tag{14}$$

where ω_p is the electron plasma frequency and V_T the thermal velocity. In writing (14) we have neglected the friction force, but it has been shown that it only slightly modifies the Jeans frequency (Bouquet, submitted to Astron. Astrophys.).

The change in sign between Eqs. (13) and (14) is crucial, and makes the gravitational system unstable for wavelengths larger than the Jeans length λ_j, defined by

$$\lambda_j = \frac{\sqrt{3}V_T}{\omega_{jo}}. \tag{15}$$

It must be pointed out that it is only in the expanding self similar system that the analysis leads to Eq. (14) and the Jeans instability has a meaning. Indeed, such an analysis supposes the existence of a steady homogeneous state with a neutralizing background. Such a state exists only in the rescaled frame and, consequently, the concept of Jeans instability and self similarly expanding universe are strongly connected, a point often insufficiently stressed.

3. Simulation

3.1 Algorithm

The one-dimensional motion of the particles (plane sheets) is calculated by an exact code similar to the one in Feix (1969): the field created by a sheet being constant and the background being uniform, the motion in the time interval between two crossings of particles can be exactly computed.

The total field is piecewise linear in space and the sheets are allowed to pass freely through each other. We calculate the time of the next crossing between a sheet and its right-hand side neighbor. We find the smallest crossing-time and advance the two particles to the position they will have at this time while the positions of all other particles are frozen at their last crossing-time. Then we calculate the new crossing-times of the three particles which have changed their right-hand side neighbor. The same process is repeated.

As units we choose: $4\pi G = 1$; superficial density equal to one; length of the system equal to the number of particles. With these choices, $\omega_{j_o} = 1$.

3.2 Initial Condition

The system is initially homogeneous: the particles are located at equal intervals in configuration space, and their velocities are randomly and uniformly distributed between the values $-V_{max}$ and V_{max}. V_{max} is much less than the velocity which makes a particle cross the whole length of the system in several units of time. We recall that this velocity is a perturbation around the expansion velocity and that the latter has been completely taken into account by the rescaling method (leading to the friction and the background force).

In order to simulate a portion of an infinite universe we will take periodic boundary conditions. To avoid unphysical changes when a particle leaves the system, we impose it to be symmetric. We thus simulate half of our symmetrical system, enclosing N particles in a box of length L with reflecting edges.

3.3 Results from Simulations

We present the results of two simulations a and b for which we have

simulation a: $N = 10000$ and $V_{max} = 10$
simulation b: $N = 10000$ and $V_{max} = 100$

Figures 1 and 2 give the evolutions of these two systems through the representations of the density profile, of the velocity distribution and of the phase space distribution for the particles. The pictures are labelled with the different times of evolution $\omega_{j_o}\hat{t} = 0, 3, 6, 8, 10$ and 12. (It might be recalled here that 'small' times correspond in fact to large expansion factors, because of the logarithmic compression of $\hat{t}$.)

These figures show that, progressively, the system breaks up into a large number of subsystems (clusters) with a uniform distribution inside the box (the clusters appear around $\omega_{j_o}\hat{t} = 5$). The larger the maximal initial velocity V_{max}, the bigger the size of these clusters, with, consequently, a smaller number of clusters. After

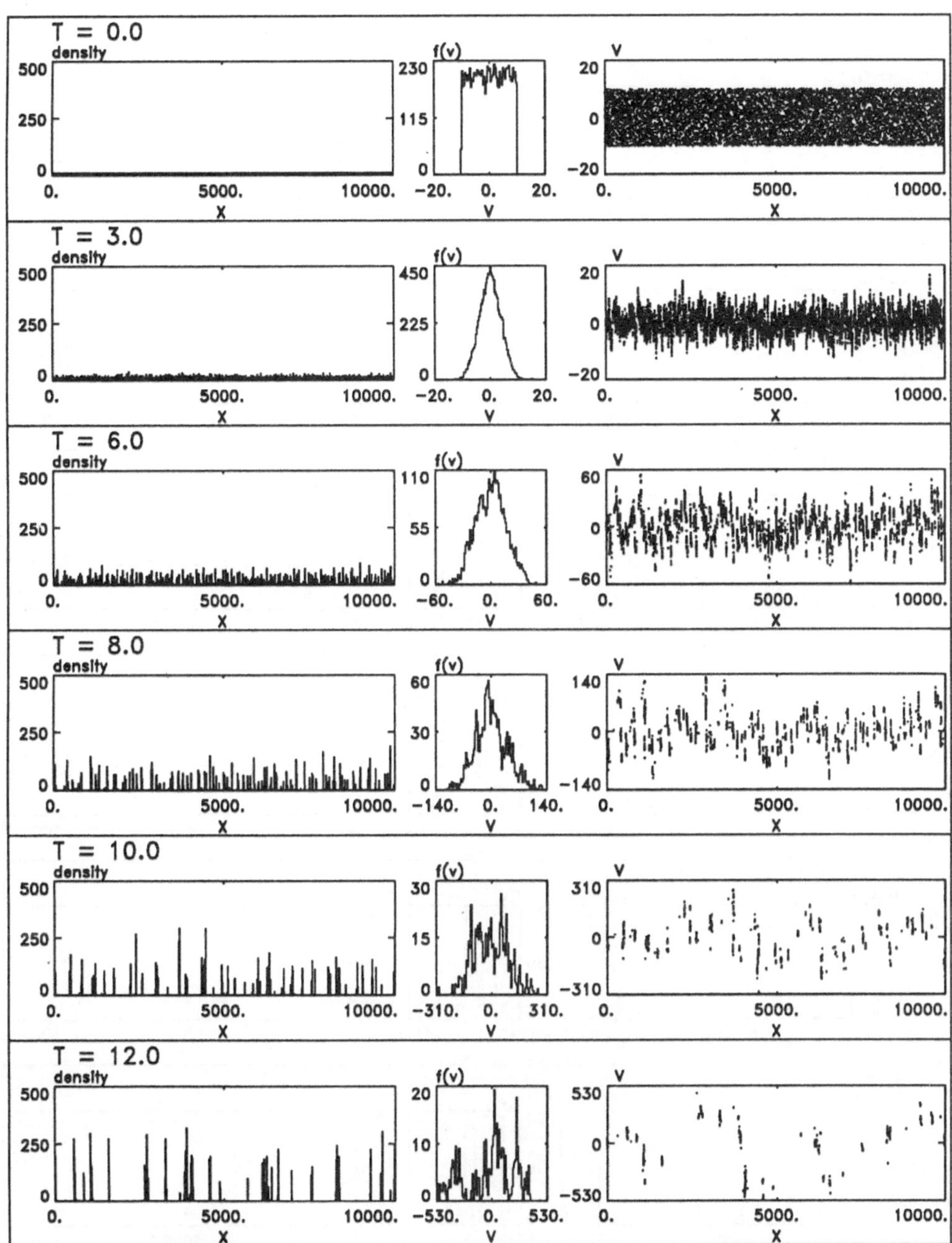

Fig. 1. Data obtained in the numerical simulation of a system of 10000 particles (stars) shown at times $T(=\omega_{j_o}\hat{t}) = 0, 3, 6, 8, 10, 12$. The initial velocity perturbation $V_{max} = 10$ (case a). Left panels show the configuration space density, with the progressive appearance of clusters of particles, clusters of clusters etc. Middle panels show the velocity distribution initially homogeneous, then maxwellianized, and becoming progressively more turbulent (velocity spikes). The right panels localize all particles in phase space, and exhibit the simultaneous processes of position clustering and velocity clumping. Velocity scales vary from panel to panel.

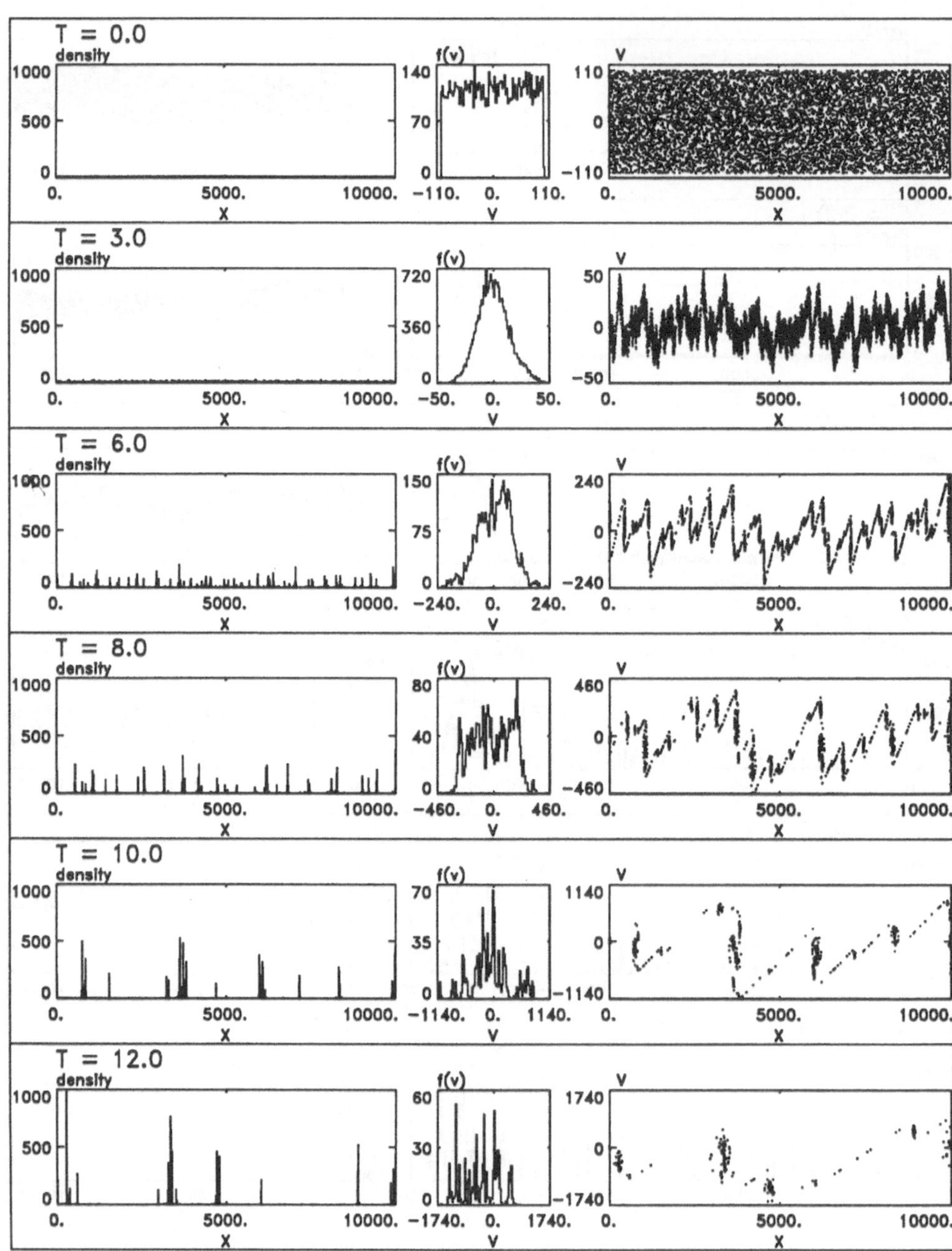

Fig. 2. Same as Fig. 1 with initial velocity perturbation $V_{max} = 100$ (case b).

this first step, the clusters themselves show a tendency to come together, forming clusters of clusters uniformly distributed inside the box. This hierarchical organization process goes on during the entire simulation. Of course the results obtained will be meaningful only if the boundary effects remain small, which is the case if the number of structures remains sufficiently large (at least bigger than 10).

Through the use of colored particles in the phase space representation, we established that, once caught by a cluster a particle remains indefinitely inside this cluster, which can, consequently, be considered as a 'superparticle'. In the same way, the superclusters formed by these superparticles keep their cohesion during the simulation.

An important difference must be pointed out between the first step in the hierarchical organization and the following ones. In the first step the number of particles inside a cluster depends on V_{max} as shown by a comparison between simulation a and b at $\omega_{jo}\hat{t} = 6$ (this number is roughly proportional to V_{max}). At the following steps, the number of 'superparticles' which cluster together to form the next structure in the hierarchy is much smaller and varies between two and five.

It should be emphasized that our approach to the problem always treats the full N-body physical problem and never, for easier or faster numerical results, merges two or more particles being very close to each other in phase space. Although this would very likely be legitimate and would certainly be very useful for much longer and higher N computations, it would cast a doubt on the physical nature of the hierarchical organization, and it would not reveal in all its details the fractal structures we describe in the next section. The obtention of this detailed precision depends crucially on the efficiency of the exact algorithm used for our simulations, in which motions at all scales are correctly computed with the maximal (64 bits) precision.

4. Fractal Nature of the Universe

4.1 Measurements of the Size of the First Generation Clusters

Although this measurement is not directly connected to the determination of a fractal dimension, we would like to mention it here. From (14) the smallest unstable wavelength initially present is the Jeans length λ_j as defined by (15). The initial perturbation of the system (consequence of its grainy structure) can be considered as a white noise. We consequently suppose that the size of the first clusters obtained will be of the Jeans length. Since the chosen units are such that $L = N$, in this hypothesis each cluster will contain roughly speaking λ_j particles.

It must be realized that a cluster is an entity which exists in phase space and the counting of the clusters is obtained by a treatment of the phase space data. We choose the following criteria: two particles belong to the same cluster if their difference in position is less than Δx and their difference in velocity less then Δv. A third particle will belong to this same cluster if its distances in position and velocity with respect to at least one of the two previous particles are less than Δx and Δv. We go on until no other particle can be associated with the cluster.

In our units $\lambda_j \sim V_{max}$ (see Eq. 15) and it is thus reasonable to take numerically $\Delta x = \Delta v$. Now if the chosen value for $\Delta x (= \Delta v)$ is too small, we get N clusters with only one particle per cluster. On the other hand if $\Delta x (= \Delta v)$ is too large, we get only one cluster with all particles in it. All the possible values for $\Delta x (= \Delta v)$ are considered but, since we do not want to consider clusters with too few particles, we count only those clusters with, at least, λ_j particles. This number increases with increasing $\Delta x (= \Delta v)$, goes through a maximum and then decreases. We take this maximum value as an indication of the number of clusters.

For a simulation with N particles, we expect to find that the number of clusters N_c is approximately

$$N_c \sim N/\lambda_j \sim N/V_{max}. \tag{16}$$

We have determined N_c for different simulations with $N = 5000$ particles and V_{max} varying from 10 to 150.

Figure 3 shows that the relation between $\ln N_c$ and $\ln V_{max}$ is a straight line of slope -1 in agreement with our assumption giving λ_j for the size of these first appearing clusters.

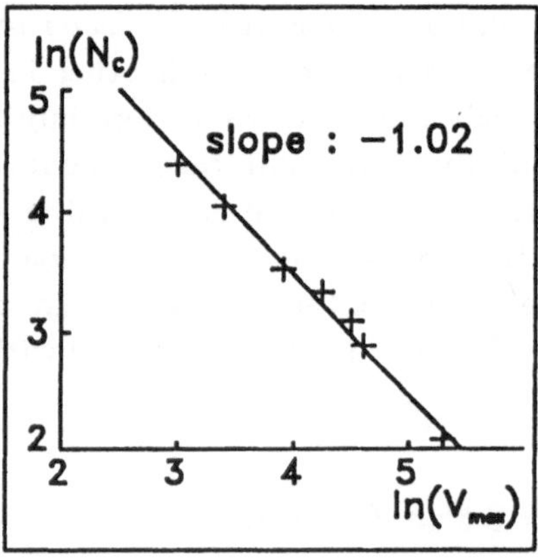

Fig. 3. Measurement of the size of the first appearing clusters.

4.2 Fractal Nature of the Clustering Process

On very first inspection of Figs. 1 and 2, the formation of clusters of particles and the evolution of these clusters into fewer 'superclusters' containing an ever increasing number of particles, are extremely reminiscent of fractal figures (Mandelbrot 1983, Barnsley et al. 1988). Indeed if one examines the density of particles in configuration space, velocity space, and their combined phase space, it is readily apparent that the hypothesized fractal nature of the clustering is most visible in configuration space, whereas, in velocity space, the most apparent features are velocity spikes without prominent fractal structure. However when observed in phase space, a configuration space cluster seems to exhibit clustering also in the second (velocity) dimension of phase space. The following analysis sheds some light on this point and qualifies more precisely these observations.

Our analysis is principally based on the simple determination of the *box counting dimension D*, where the number of boxes containing one or more particles is plotted on a log-log scale versus the dimension of the box and the slope of the linear best fit determines the fractal dimension (e.g. Falconer 1985, Feder 1988). As pointed out at the end of Sect. 3, the very detailed and precise knowledge of our particle distribution allows the computation of the fractal dimensions over the largest possible span of distances, from a sizeable part of the whole system to the space occupied by two neighboring particles, which, in our simulations represents over 3 orders of magnitude.

4.2.1 Configuration Space

In Figs. 4 and 5, we show the box counting plots for density in configuration space for simulations a and b presented above. They clearly show that the homogeneous density ($D = 1$) transforms progressively around $\omega_{j_o}\hat{t} = 4$ (case a) and $\omega_{j_o}\hat{t} = 5$ (case b) in a fractally clustered configuration with $D = 0.55 \pm 0.02$ (sampled between $6 \leq \omega_{j_o}\hat{t} \leq 14$) in case a and $D = 0.57 \pm 0.02$ (sampled between $7 \leq \omega_{j_o}\hat{t} \leq 14$) in case b.

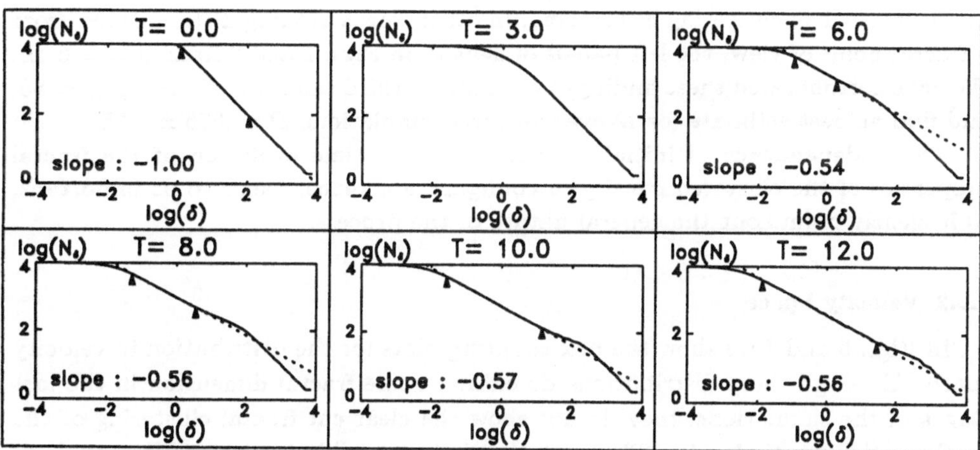

Fig. 4. Box counting determination of the fractal dimension for the densities at times $T(= \omega_{j_o}\hat{t}) = 0, 3, 6, 8, 10, 12$ in simulation a (left panels in Fig. 1). The slope of the straight line (dashed) giving the best fit for all data between triangles, is indicated in each panel. δ = box dimension; N_δ = number of non empty boxes.

This fractal clustering progressively extends from dimensions of order unity to both smaller and larger scales. The essential features of the process are:

- the two cases are fractally similar

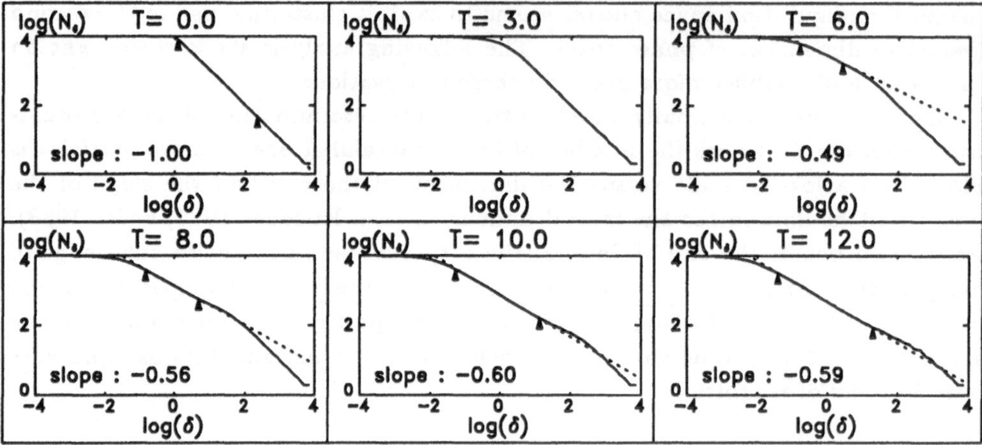

Fig. 5. Same as Fig. 4 for simulation b corresponding to the left panels in Fig. 2.

- the fractality of the structure, when formed, remains largely independent of time

Indeed if case a and b exhibit differences in initial stage and speed of clustering, they are however indistinguishable in their fractal nature and dimension, and this at all times when clustering has started. In other words nothing distinguishes, from a fractal point of view, the left panels of density in Figs. 1 and 2 from $\omega_{j_o}\hat{t} = 6$ on. We have corroborated these findings by running a third simulation with $V_{max} = 50$, and find as best estimate for D over all three simulations $D = .575 \pm .035$.

This independence of initial conditions and of time evolution of the fractal properties of the clustered density in configuration space, seems to us important, as it clearly points out the general nature of the process.

4.2.2 Velocity Space

In Figs. 6 and 7 we show the box counting plots for the distribution in velocity space. Although these distributions do exhibit some fractal dimension in the last stages of the simulations, they do not show the clear cut fractal clustering of the configuration space density. The evolution is best visible in the middle panels of Figs. 1 and 2. The distribution is very quickly thermalized and, then, becomes increasingly random, developing large velocity spikes, which go on increasing. The voids thus produced in the distribution are mostly responsible for the less than one fractal dimension, and for the difference between cases a and b. Consequently, the value of the fractal dimension should be taken with caution as these voids could be, even if very partially, filled in simulations with more particles, and thus change the box counting. The understanding and the structure of the velocity spikes must be done in correlation with the density clusters, by the analysis of the full phase space distribution fractal structure.

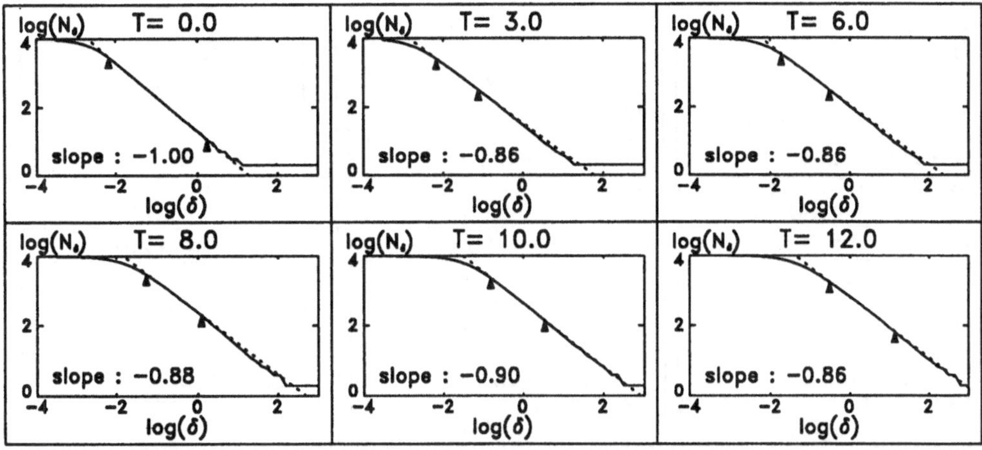

Fig. 6. Box counting determination of the fractal dimension for the velocities at times $T(=\omega_{j_0}\hat{t}) = 0, 3, 6, 8, 10, 12$ in simulation a (middle panels in Fig. 1). The slope of the straight line (dashed) giving the best fit for all data between triangles, is indicated in each panel. δ = box dimension; N_δ = number of non-empty boxes.

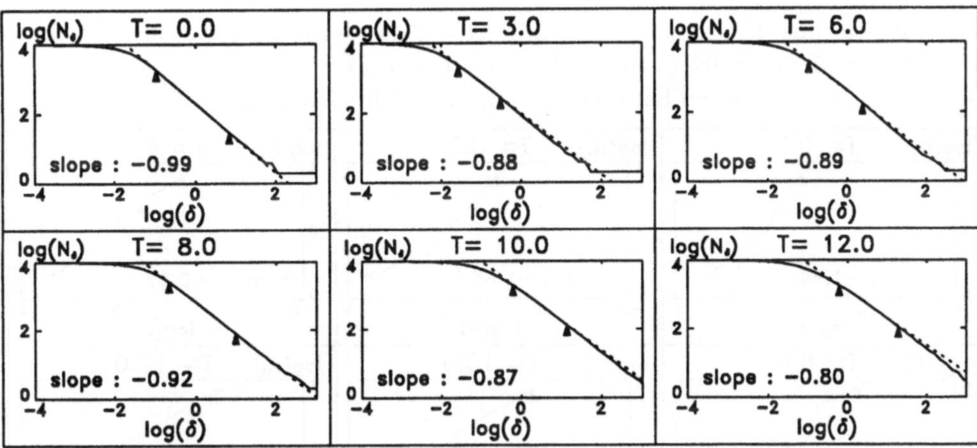

Fig. 7. Same as Fig. 6 for simulation b corresponding to the middle panels in Fig. 2.

4.2.3 Phase Space

The phase space evolution is initially characterized by a gaussian randomization process of the velocities, physically due to friction, followed by the clustering process in the x coordinate, each cluster (supercluster) seeing the velocities of its constituent particles progressively clumped within a limited range. From a fractal point of view, this induces, as is shown in Figs. 8 and 9, a very rapid initial decline

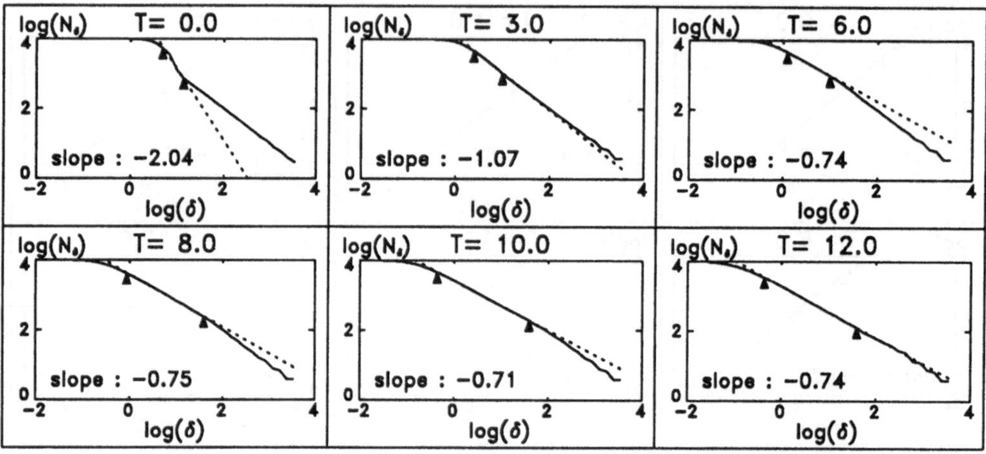

Fig. 8. Box counting determination of the fractal dimension for the phase space distribution at times $T(=\omega_{jo}\hat{t}) = 0, 3, 6, 8, 10, 12$ in simulation a (right panels in Fig. 1). The slope of the straight line (dashed) giving the best fit for all data between triangles, is indicated in each panel. δ = box dimension; N_δ = number of non empty boxes.

of the fractal box dimension from 2 at $\omega_{j_o}\hat{t} = 0$ to reach, more slowly after that, a value $D = 0.73 \pm 0.02$ (sampled between $6 \leq \omega_{j_o}\hat{t} \leq 14$) in simulation a and $D = 0.77 \pm 0.02$ (sampled between $7 \leq \omega_{j_o}\hat{t} \leq 14$) in simulation b.

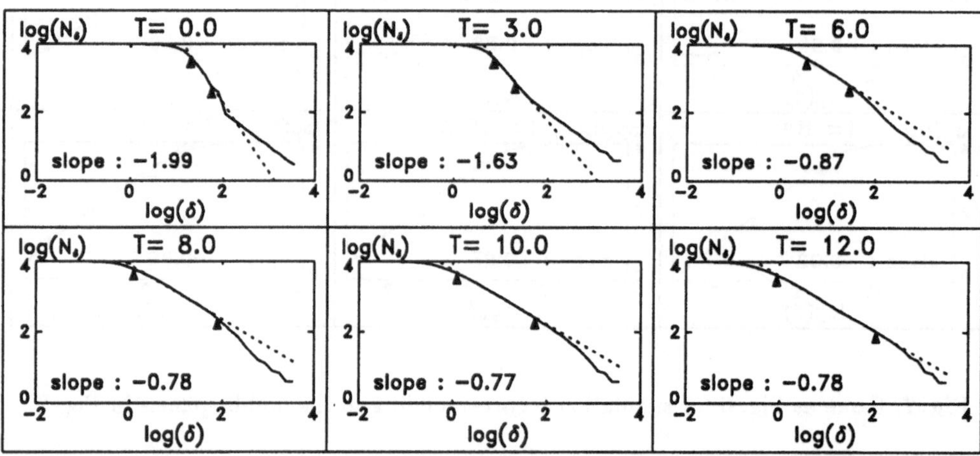

Fig. 9. Same as Fig. 8 for simulation b corresponding to the right panels in Fig. 2.

This indicates a very strong fractal structure, which cannot result from the sole superposition of the analysis in the separate x and v dimensions. This also clearly points out that phase space is the adequate space to study turbulence in general and its fractal structure in particular for these large gravitational structures.

To try to better illustrate the fractal nature of the velocity clumping, we show in Figs. 10 and 11 a series of successive enlargements of the phase space distribution at time $\omega_{j_0}\hat{t} = 13$ for simulation a and b respectively; we have also represented the configuration space density to better distinguish the reality of velocity clumping without regard to density clustering. This illustration helps to visualize the phenomenon, which explains at least qualitatively the complex fractal structure and the difference it induces between the box counting dimensions in configuration space, velocity space and phase space. Notice also the richness of the fractal structure of the density (left panels in Figs. 10 and 11).

4.3 Geometrical and Physical Fractal

To conclude our analysis of fractal structures clearly exhibited by our results, which are derived from a model based on physical laws, we should emphasize that similar results (hierarchical organization exhibiting the same fractal dimension) can be obtained from other models, a priori completely decoupled from physics. Indeed it is possible to mimic the configuration space in the following way: we build a uniform distribution of P points. The distance between two points is K. At each point we put a particle. Next with each particle, we associate $2M$ particles (M on each side) at equal distance αK one from each other, with $\alpha < 1$ (i.e. with respect to the 'main' particles, the 'satellite' particle have positions $j\alpha$ with j running from $-M$ to M), and we repeat the process. With the above definitions, the fractal dimension is

$$\beta = -\frac{\log(2M+1)}{\log\alpha}. \tag{17}$$

By a proper choice of α and M it is possible to find a large number of systems with the same geometrical properties as those found in our N body problem. Each of the P points (with their uniform distribution) stands for the biggest coarse grained observed structure (at a given time). Going to a finer resolution we see that this structure is formed of $2M+1$ smaller structures and so on. Examples and results from this type of model building are presently under study.

5. Conclusion

In this paper, we carried on the numerical simulation on an expanding gravitational system under the action of the sole Newtonian force. The system is very close to the self similar Hubble expansion, which is analytically treated through the rescaling group. The system is initially homogeneous in space and is uniformly distributed between $-V_{max}$ and V_{max} in velocity space. A white noise is naturally induced by the velocity perturbation at time $t = 0$ and triggers instabilities over the complete range of wavelengths, which in turn produce the breaking up of the system into subsystems of Jeans size.

The results of two simulations differing by the intensity of the initial perturbation have been presented. The hierarchical structure of particles aggregating in

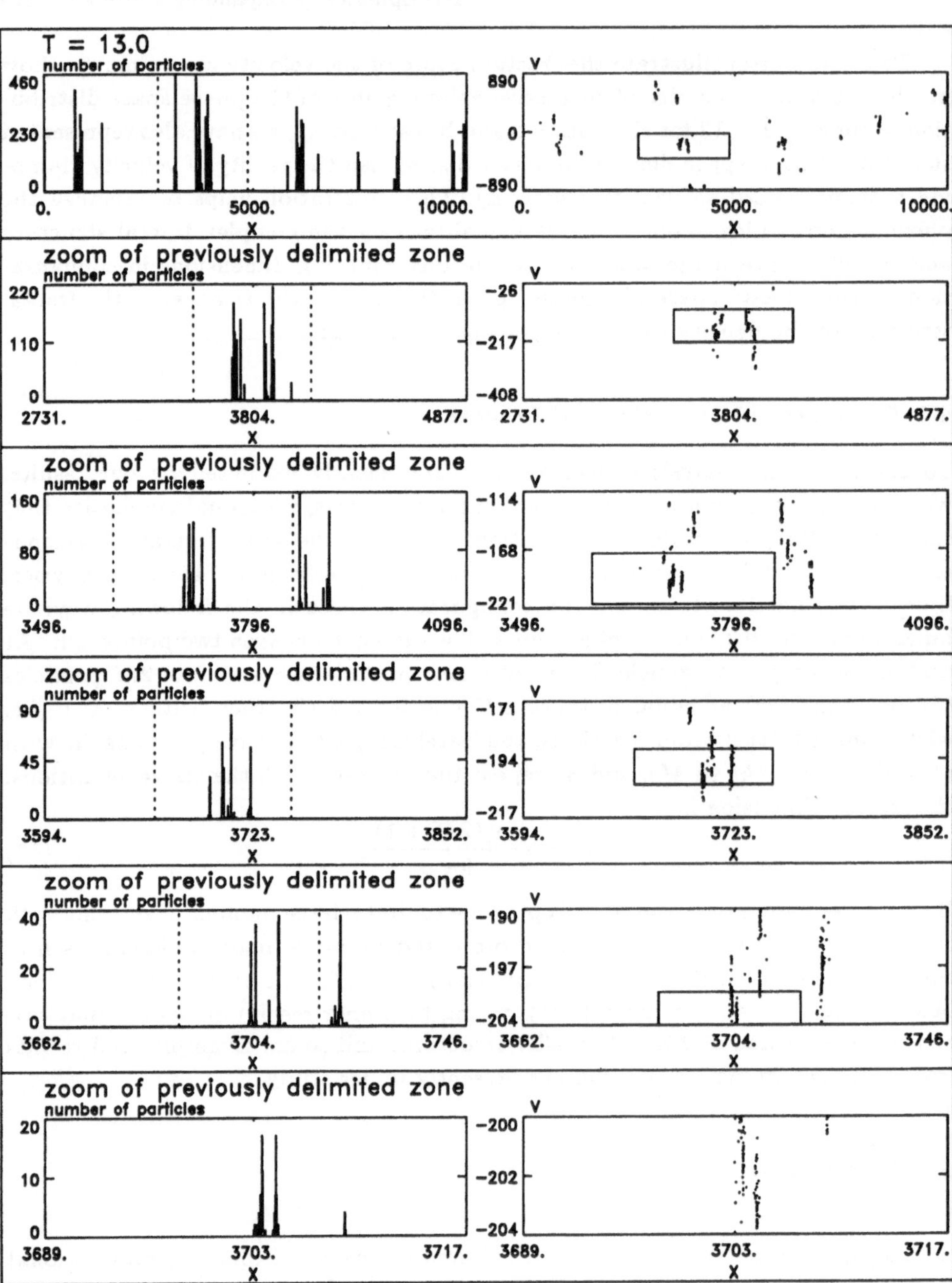

Fig. 10. Phase space distribution (right panels) and corresponding histogram of number of particles in x (left panels) for simulation a at time $T(= \omega_{j_o}\hat{t}) = 13$. Progressively enlarged views illustrate with increasing detail the complementary fractal structures in both coordinates of phase space. Each small rectangle in phase space is enlarged in the picture exactly below; the same area is delimited in the corresponding left panel by dotted lines. Absolute x and v coordinates are indicated in each panel. Histograms of number of particles are drawn using 1000 equal subintervals in shown interval.

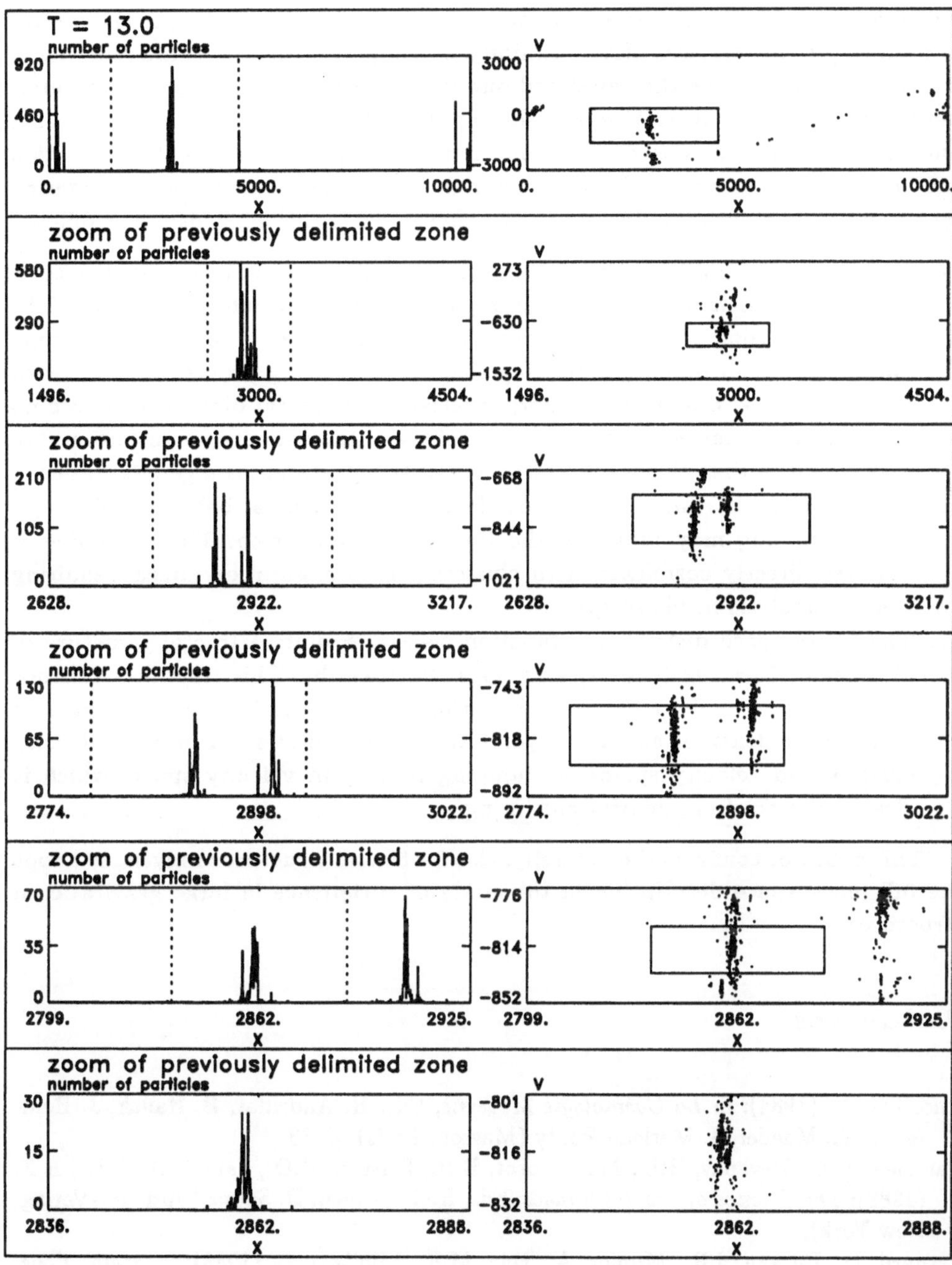

Fig. 11. Same as Fig. 10 for simulation b.

clusters, which behave themselves as true superparticles and aggregate themselves in even larger clusters, etc. clearly exhibits, on simple inspection, characteristics of fractal nature. We have thus analyzed our results with the well known box counting technique, particularly adequate for our simulation which is numerically exact and where the individuality of all 10000 particles is kept throughout the simulation (no 'numerical clustering'); this allows our analysis to proceed over scales differing by 3 to 4 orders of magnitude. The main conclusions of this analysis are as follows:

a. The density in configuration space unequivocally exhibits a fractal structure, which the system acquires rather rapidly (starting around $\omega_{j_o}\hat{t} = 5$). The fractal dimension of this structure is .575 ± .035, and does not evolve in time, once it has been built up, meaning that the clustering process goes on with identical fractal dimension (superparticles). This fractal dimension shows no clear cut dependence on the amplitude of the initial velocity perturbation. We may conclude that this fractal structure of the clustering is a general process.
b. The velocity distribution does not clearly exhibit fractal behavior, although some clumping may influence the fractal box dimension. This clumping is however directly correlated with clustering in configuration space, requiring thus an analysis in phase space.
c. The phase space distribution function clearly exhibits a fractal structure of dimension .73 ± .02 (case a) and .77 ± .02 (case b). This fractality can not be totally ascribed to the sole clustering in configuration space. Images of phase space densities exhibit, beyond the obvious correlation between density clusters and velocity spikes, a clumping process in velocity space, which is clearly distinct from density clustering.

This behavior confirms the necessity, also for fractal structure analysis, to adopt the full phase-space description in the study of turbulence in large gravitational structures.

References

Andrillat, H. (1984): in *La Cosmologie Moderne*, Eds. H. Andrillat, B. Hauck, J. Heidmann, A. Maeder, J. Merleau-Ponty (Masson, Paris), p. 33.

Barnsley, M.F., Devaney, R.L., Mandelbrot, B.B., Peitgen, H.O., Saupe, D., Voss, R.F. (1988): *The Science of Fractal Images*, Eds. H.O. Peitgen, D. Saupe (Springer-Verlag, New York).

Besnard, D., Burgan, J.R., Munier, A., Feix, M.R., Fijalkow, E. (1983): *J. Math. Phys.* **24**, 1123.

Bouquet, S., Feix, M.R., Fijalkow, E., Munier, A. (1985): *Astrophys. J.* **293**, 494.

Doroshkevich, A.G., Kotok, E.V., Novikov, I.D., Polyudov, A.N., Shandarin, S.F., Sigov, Yu.S. (1980): *Monthly Not. Roy. Astron. Soc.* **192**, 321.

Falconer, K.J. (1985): *the Geometry of Fractal Sets* (Cambridge Univ. Press, Cambridge).

Feder, J. (1988): *Fractals* (Plenum Press, New York).

Feix, M.R. (1969): in *Non–linear Effects in Plasmas*, Eds. G. Kalman, M.R. Feix (Gordon and Breach, New York), p. 151.

Feix, M.R. (1978): in *Strongly Coupled Plasmas*, Eds. G. Kalman, P. Carini (Plenum Press, New York), p. 499.

Mandelbrot, B.B. (1983): *The Fractal Geometry of Nature* (Freeman, New York).

Munier, A., Feix, M.R., Fijalkow, E., Burgan, J.R., Gutierrez, J. (1979): *Astron. Astrophys.* **78**, 65.

Peebles, P.J.E. (1980): *The Large-Scale Structure of the Universe* (Princeton Univ. Press, Princeton).

Rouet, J.L., Feix, M.R., Navet, M. (1990): in *Fractals in Astronomy*, Ed. A. Heck, *Vistas Astron.* **33**, 357.

Soneira, R.M., Peebles, P.J.E. (1977): *Astrophys. J.* **211**, 1.

The Fractal Structure of the Quantum Space-Time

Laurent Nottale

C.N.R.S., Département d'Astrophysique Extragalactique et de Cosmologie, Observatoire de Meudon, 5 Place Jules Janssen, F–92195 Meudon Cedex, France

Abstract: We report here on the present state of an attempt at understanding microphysics by basing ourselves on a 'principle of scale relativity', according to which the laws of physics should apply to systems of reference whatever their scale. The continuity but non differentiability of quantum mechanical particle paths, the occurrence of infinities in quantum field theories and the universal length and time scale dependence of measurement results implied by Heisenberg's relations have led us, among other arguments, to suggest the achievement of such a principle by using the mathematical tool of fractals (Nottale 1989). The concept of a continuous and self-avoiding fractal space-time is worked out: arguments are given for generally describing them as families of Riemannian space-times whose curvature tends to infinity when scale approaches zero. We recall some basic results already obtained in this quest: the fractal dimension of quantum trajectories jumps from 1 to 2 for all 4 space-time coordinates, with the fractal/non fractal transition occurring around the de Broglie's length and time; conversely the Heisenberg relations may be obtained as a consequence of assumed fractal structures; point particles following curves of fractal dimension 2 are naturally endowed with a spin. We develop the interpretation of the wave-particle duality as a property of families of geodesical lines in a fractal space-time and examplify it with the Young's hole and Einstein-Podolsky-Rosen experiments. Finally some possible consequences concerning gravitation are recalled, with the suggestion that Newton's law may break down for active gravitational masses smaller than the Planck mass.

1. Introduction

The present contribution may at first appear as at variance with the theme of this book, fractals in astronomy. Indeed it reports on the present state and on the progress of an attempt at understanding the quantum behaviour of the microphysical world in terms of a postulated underlying fractal space-time. However, even if astrophysical considerations are not explicitely present in this work (except concerning the suggestion of a new space experiment on the law of gravitation, see

Sect. 6), the links of astrophysics and microphysics are now becoming so tight, e.g. in the domains of compact relativistic objects, of possible dark matter astroparticles and mainly of the primeval universe, that such interrogations, even if they are at present relevant of theoretical physics, are expected to find in astrophysics one of their best domain of application.

Our results are based upon an analysis of what appears as remaining inconsistencies and incompleteness in the present state of fundamental physics. We give here a summary of the principles to which we have been led in our attempt at overcoming these problems, then recall some of our main results, which are fully described in Nottale (1989), and also develop farther some aspects of the formalism and its consequences.

The first remark is that, following the Galileo/Mach/Einstein analysis of motion relativity, the non absolute character of space and space-time appears as an inescapable conclusion (Einstein 1916): the geometry of space-time should depend on its material and energetic content. However present quantum physics assumes the microphysical space-time to be Minkowskian, i.e. absolute, while the fundamental behaviour and properties of quantum objects are known to be radically at variance with classical properties, from which the Minkowskian space-time was yet derived. The absolute character of the microphysics space–time in present quantum theory is incompatible with Mach emphasis on the relativity of *all* motions, which should be applied to the microphysical domain as well.

Following the emergence of general relativity, physicists have become accustomed to identify and connect questions about space-time uniquely to gravitation. But what the general theory of relativity actually tells us is that gravitation is to be identified to the manifestations of *curvature* The geometry of space-time at the microscopic level may after all be different from Minkowskian or Riemannian, and if generalized geometric properties do exist in nature, they should manifest themselves through a non classical physical behaviour.

One may also remark that the flatness of space-time which is observed with a high precision at our scale, and even at scales accessible with microscopes, has been extrapolated (or, strictly, interpolated) to quantum scales. This is the main observational argument for the use of a Minkowski space-time, but it clearly fails precisely in the quantum domain (i.e. at scales smaller than the de Broglie wavelength of a system) inside which classical measurements are no more allowed by Heisenberg's relations. So the situation is reversed compared to general relativity. Here the constraint is to get back flatness at the larger (relative) scale, while in general relativity the equivalence principle ensures local flatness, and curvature is found at larger scales.

Einstein's principle of general relativity ('the laws of physics should apply to systems of reference in any state of motion') is in fact still unachieved. It can not, in particular, be applied under its present form to quantum motion. As remarked by Feynman, 'typical paths of a quantum-mechanical particle are highly irregular on a fine scale. Thus, although a mean velocity can be defined, no mean-square velocity exists at any point. In other words, the paths are non differentiable' (Feynman and Hibbs 1965). This means that the state of motion of a quantum-mechanical particle can be reduced neither to uniform, nor to uniformly accelerated, nor even to a description involving higher derivative of coordinates. Einstein indeed realized that giving up differentiability could be the price to be paid in order to apply the principle of relativity to microphysics. He writes in a letter to Pauli (1948): '...I am a fierce supporter, not of differential equations, but of the principle of general relativity, whose heuristic strength is essential to us' (Einstein 1948).

We seem to be faced with a situation in which we are not allowed to admit a priori that microphysics space-time is flat, while we also know that any attempt based on curvature or torsion and more generally on differentiable geometry is expected to fail. So we are led to ask ourselves which are the property (or properties) of the quantum world which are both irreducible to classical and relevant of becoming the basic stone of a new geometrical and spatio-temporal attempt. It is clear that only those properties of objects or physical phenomena which are *universal* may be transferred to becoming properties of space-time itself. It was precisely by demonstrating that the Lorentz transform applied not only to radiation, but also to matter, and thus was universal, that Einstein constructed the special theory of relativity and that Minkowski was able to substitute the concept of space-time to those of independent space and time. It was thanks to one of the main consequences of the equivalence principle, i.e. that any physical being, either matter or radiation, describes bent trajectories in a field of gravitation (Einstein 1907), that this universal curvature of trajectories can be attributed to space-time itself.

In the quantum world, there are indeed two laws which more than sixty years of experimental and theoretical work have definitively demonstrated to be universal, i.e. applying to any quantum object or system: these are (i) the wave nature of physical objects and systems, as described by the de Broglie-Einstein relations; (ii) the Heisenberg relations.

De Broglie's relations play a leading role in our approach, because one may demonstrate that the irregularity and non differentiability of quantum paths, as pointed out by Feynman, are found only at resolution smaller than the de Broglie wave-length, which thus defines a quantum-classical transition (Abbott and Wise 1981; Nottale 1989). Concerning Heisenberg's relations, they may be viewed as describing a fundamental and universal dependence of physical laws on scale (or equivalently on resolution).

We may now remark that the consequences of this radically new behaviour of the quantum world relative to the classical one have still not been fully drawn. Though an essential part of the quantum theory through the so-called measurement theory, this scale dependence is not included into the laws of physics themselves, contrarily to what one may expect from our analysis. Indeed, it is clear that a set of physical measurements takes its complete physical sense, even in the classical domain, only when the measurement resolutions or 'errors' have been specified. In the quantum domain, the result of a momentum measurement explicitely depends, although in a statistical manner, on the spatial resolution, and the result of an energy measurement depends on the temporal resolution with which the measurement has been performed.

So we have suggested (Nottale 1988a, 1989) that this fundamental scale dependence of physics is relevant, as motion does, to a relativity theory. Our proposal is to introduce explicitely the resolution in physical laws as a *state of scale* of the coordinate system, in the same way as velocity and acceleration describe its state of motion. The axes of such a generalized coordinate system can be viewed as endowed with thickness. In such a frame of thinking, one would require general covariance of physical equations, not only on motion, but also on scale, and the general principle of relativity could be still generalized under the working program: '*the laws of physics should apply to systems of references, whatever their states, either of motion or of scale*'.

The hereabove generalization in the definition of coordinate systems, once assumed universal for a consistent description of physical laws, immediately implies a generalization of the nature of space-time itself. We postulate that the scale dependence of physics in the quantum domain takes its origin in an intrinsic dependence of space-time geometry on resolution, and then attempts at understanding the quantum behaviour itself in such terms. This may be achieved from a generalized metric element in which the metric potentials become explicit functions of resolution:

$$ds^2 = g_{\mu\nu}(x^\lambda, \Delta x^\lambda)dx^\mu dx^\nu \tag{1}$$

The achievement of the hereabove working hypothesis, whose physical sense will be enlighted hereafter, needs the use of an adequate geometrical tool. We have suggested that the concept of a *continuous and self-avoiding fractal space-time* might be such a tool (Nottale 1989). Such a space-time would not be absolute, but instead its geometrical structure would depend on its material and energetic content: for example it will be shown hereafter that it would be subjected to a fractal-non fractal transition around the de Broglie length and time, which are themselves functions of energy and momentum.

2. Why and Which Fractals?

The suggestion that the quantum space-time possesses fractal structure is supported by a lot of converging arguments. First of all, fractals are characterized by an effective and explicit dependence on resolution. Covering a fractal domain of topological dimension T and fractal dimension $D > T$ by balls of radius Δx yields a T-hypervolume measure which diverges when Δx approaches zero as:

$$V(x, \Delta x) = \xi(x, \Delta x)\left(\frac{\lambda}{\Delta x}\right)^{D-T} \tag{2}$$

where $\xi(x, \Delta x)$ is a finite fractal function and λ is a fractal / non fractal scale transition. More will be said in Sect. 3 about what is meant by 'fractal function' and about how they emerge from scale relativistic requirement.

Second, fractals are also characterized by their non-differentiability, while it had be realized by Einstein that differentiability and general relativity might well be in conflict in the microphysical domain, and by Feynman that the quantum 'trajectories' of particles can be described as continuous but non differentiable curves (see the hereabove quotations). Note that the choice of a continuous fractal space is less radical than the quantum gravity proposals, assumed to hold at the Planck length scale, that the topology of space-time itself could be broken through a foam-like structure (Fuller and Wheeler 1962; Hawkings 1978): in this last hypothesis it is continuity itself which would be given up instead of only differentiability.

The third point is that fractals contain infinities, which may be renormalized in a natural way (Nottale and Schneider 1984; Nottale 1989). The correspondence between the renormalization group methods, particularly efficient in the domain of the asymptotic behaviour of quantum field theory, and fractals, has already been pointed out by Callan, as quoted by Mandelbrot (1982a) and more recently by Sagdeev et al. (1988). In fact the fractal approach may come as a completion of the renormalization group theory. Indeed the renormalisation group is, strictly, a semi-group (Le bellac 1988; Wilson 1979), since its basic transformation, integration on small scales to get a larger scale, has no inverse in the present theory. Such an inverse transformation would mean to be able to get information on the smaller scale from what is known at the larger. *This is precisely what a fractal generator makes.*

Our working hypothesis is that the quantum space-time is a fractal self-avoiding continuum whose geodesics define the trajectories of free particles. The continuum assumption is a conservative choice allowing to keep a field approach and representation, while the self-avoidance assumption is a necessary condition for such a geodesical interpretation. But because of the non-differentiability of a fractal space, the properties of geodesics will be fundamentally different from those of standard spaces (see Nottale 1989 and Sect. 5 hereafter).

The fact that fractal trajectories run backward relatively to classical coordinates also implies a loss of information in the projection from the intrinsic (fractal) to the classical coordinates, which are defined as the result of a smoothing out of the fractal coordinates with balls larger than the fractal/non fractal transition λ. We have proposed to describe this behaviour by a 'fractal derivative' (Nottale 1988a, 1989):

$$\frac{d\bar{s}}{dx} = \sum_i \frac{ds_i}{dx} = \xi'(x, \Delta x)\left(\frac{\lambda}{\Delta x}\right)^{D-T} \tag{3}$$

the sum being performed on all intrinsic fractal coordinate intervals found inside the classical interval dx . The power $D-T$ is recovered because both the topological dimension and the fractal dimension (nearly everywhere, Mandelbrot 1982b) are decreased by 1 in the projection.

3. Towards an Intrinsic Definition of Fractal Geometry

The idea of a fractal space (or space-time) stands as a basic stone of the present approach and should now be made precise. We will attempt in the present section at proceeding further with respect to the state of the question as it was left in Nottale (1989). Up to now (and to our knowledge), even if the words 'fractal geometry' have been extensively used and are in some sense justified, the mathematical domain of fractals has not yet reached a status comparable to Euclidean geometry or Riemannian geometry. Fractals in the present literature are either a characterization of some properties of some phenomena in nature, or objects or sets most of the time seen as embedded into an underlying Euclidean space.

Our approach here is different. The discovery by Gauss of a curved geometry which definitely proved the fifth axiom of Euclid to be independent from the remaining ones, was made effective when he became able to characterize a surface by totally *intrinsic* methods. Thus not only already known surfaces like spheres could be defined and seen from the inside, but moreover this allowed to prove the logical existence of surfaces which cannot be embedded into Euclidean space, like the hyperbolic one, and which deserved the name of new geometry. Once the 'Pandora box' opened, it was once again Gauss' merit to partially close it by restricting himself to spaces which remain *locally Euclidean*, an hypothesis which was kept by Riemann in his generalization to any topological dimension. One knows the fundamental part played by Gauss' hypothesis in the use of Riemannian geometry as a mathematical tool for the achievement of the general theory of relativity: it is the basis for the mathematical transcription of the equivalence principle.

Now with fractals the Pandora box is opened once again. Fractals may be seen as a dismissal of Gauss' hypothesis: instead of getting flatness and rectifiability back at a small scale, on the contrary the never ending occurrence of new details as scale gets smaller and smaller implies that the curvature of a fractal space increases up to infinity as resolution Δx approaches zero (Nottale 1989). Fractal surfaces may be built which are flat at the larger scale (once the fractal-non fractal transition crossed), and whose curvature at the infinitesimal level is everywhere infinite, but with a fractal distribution of negative and positive signs.

How can we actually build such a new tool? We have proposed (Nottale 1989) to define a fractal space-time as a family of Riemannian space-times. Let us make precise and justify this choice. Recall first that one of the best ways to work on fractals and to deal in a consistent mathematical way with their non-differentiability and their infinities is to use Non Standard Analysis (Nottale and Schneider 1984). We recall briefly that Non Standard Analysis is a mathematical frame worked out by Robinson (1961) and later axiomatized by Nelson (1977) in which one may explicitely work with infinitesimal and infinite numbers. An infinite number has the property to be larger than any integer, and an infinitesimal to be smaller than the inverse of any integer. An important theorem is that any finite non-standard number may be decomposed in a *unique* way into a standard number and an infinitesimal one: $a = x + dx, a \in {}^*\boldsymbol{R}, x \in \boldsymbol{R}, dx \in \boldsymbol{I}$, the set of infinitesimals. This is one aspect of the axiom of standardisation, which allows to relate non-standard objects to their standard counterparts.

In Nottale and Schneider (1984), we have generalized the construction of fractals by first defining a non-standard fractal F_ω from applying a generator ω times (thus getting a scale factor $q^{-\omega}$), ω being an infinite integer, then by taking its standard part: $F = st(F_\omega)$. While strictly F is non differentiable, one may replace F_ω by some non-standard set G_ω whose standard part is always F, but for which a kind of derivative may be defined (Nottale and Schneider 1984): to simplify the argument, it is equivalent to saying that the same standard F may be obtained by applying an infinite reduction factor q^ω to some differentiable set (itself infinite).

The interest of this approach is that, once seen under such an infinite magnifying glass, a fractal space recovers the usual properties of differential geometry; we may apply to it Gauss' hypothesis and assume that it becomes locally Euclidean, i.e. flat at a scale $\ll q^{-\omega}$. In other words one may define a fractal continuous space-time as becoming Riemannian once magnified by a factor q^ω.

There is a complementary and equivalent approach to the same problem. Consider the successive approximations F_n of a fractal F. They may be obtained by several ways. The first one is by successive application of a generator G. In case of perfect self-similarity, this generator is unique and independent of scale: $F_n = G^{(n)}(F_0)$. But the generator may also change with scale: $F_n = G_n\Big[G_{n-1}\big[...[G_1(F_0)]\big]\Big]$. Conversely one may start from the fractal set F itself and smooth it out with balls of characteristic size ε, thus defining a continuous approximation $F(\varepsilon)$:

$$F(x,\varepsilon) = \int \phi_\varepsilon(x,y)\ \ F(y)\,dy \tag{4}$$

where ϕ_ε is a smoothing function (e.g. a loophole with halfwidth ε and center x, or a Gaussian function with standard error $\approx \varepsilon$, ...). Assume that the scale factor in the first construction is q, then the resolution on F_n is q^{-n}, and we may require the equivalence of both methods with $\varepsilon = q^{-n}$. However this equivalence should be demanded only to some given uncertainty which is precisely of the order of ε. So, in the same way as, in the non-standard construction, we were able to get the same fractal as the standard part of various non-standard sets, the same fractal F may also be obtained as the limit of various families of sets $F(\varepsilon)$ which differ one from each other only by ε, *for all values of* ε. Then one is free to choose differentiable sets for the various $F(\varepsilon)$. For example the approximations defined in (4) are clearly differentiable whatever $\varepsilon \neq 0$, even if their limit $F = F(0)$ is not.

Let us examplify these ideas on a simple example. Consider the well known von Koch curve, whose generator is made of 4 segments of length $1/3$. Each successive approximation F_n in the usual construction is non differentiable on the points which connect the segments of length 3^{-n}. This construction may anyway be considered as a point by point building of the fractal curve, the various segments connecting these points being there only to ensure continuity, but having no meaning for the final fractal. So one may replace each F_n by differentiable curves F'_n passing through these building points and get at the end the same fractal. A possible method to do that would be for example to draw a polynomial $y = a_k(x)^k$ (Einstein notation) through the building points (x^i, y^i); the coefficients a_k are solution of the equation $M^{ik}a_k = y^i$ where $M^{ik} = (x^i)^k$, i.e. $a_k = M_{kj}y^j$, with $M^{ik}M_{kj} = \delta^i_j$.

This also leads to a definition for *fractal functions*. We shall define a fractal function as a family of differentiable curves, modulo some equivalence relation:

$$y = f(x;\varepsilon) \tag{5}$$

$$f \equiv g \Leftrightarrow \forall\varepsilon, \forall x, \quad |f(x;\varepsilon) - g(x;\varepsilon)| < \varepsilon \tag{6}$$

This definition corresponds to ε being a resolution of the y coordinate. The transitivity of the equivalence relation is ensured by the fact that the inegality must be true whatever the value of ε. If for a given value of ε, one has three functions such that $|f - g| < \varepsilon$, $|g - h| < \varepsilon$ but $|f - h| > \varepsilon$, one expects that for $\varepsilon' \ll \varepsilon$, either $|f - g| > \varepsilon'$ or $|g - h| > \varepsilon'$. This definition of the equivalence relation may be made more general. Assume that some distance $\mathcal{D}[x, y; x_1, y_1]$ has been defined in the plane, we may set:

$$f \equiv g \Leftrightarrow \forall\varepsilon, \forall x, \quad \exists x_1, \quad \mathcal{D}[x, f(x;\varepsilon); x_1, g(x_1;\varepsilon)] < \varepsilon \tag{7}$$

There is a still more economical definition, which allows to refer ε only to the x axis (and so could be more adapted to some physical problems):

$$f \equiv g \Leftrightarrow \forall\varepsilon, \forall x, \ \exists x_1, \ |x - x_1| < \varepsilon, \ f(x;\varepsilon) = g(x_1;\varepsilon) \text{ or } f(x_1;\varepsilon) = g(x;\varepsilon) \tag{8}$$

But the hereabove definitions may apply to standard functions as well, i.e. the fractal behaviour of the family of functions should now be characterized. Let us add two additional relations:

$$\forall\varepsilon, \quad \forall\eta < \varepsilon, \quad \exists\phi_{\varepsilon\eta}, \ f(x;\varepsilon) = \int \phi_{\varepsilon\eta}(x, y) f(y;\eta)\, dy \tag{9}$$

$$\frac{df}{dx}(x;\varepsilon) = \varphi(x) + \xi(x;\varepsilon)\left(\frac{\lambda}{\varepsilon}\right)^{\delta} \tag{10}$$

The first relation expresses the fact that the function at resolution ε is a smoothed version of the function at a smaller resolution η. This is a more local (in terms of resolution) version of (4), in which each approximation is defined from smoothing the limit fractal itself $f(x, 0)$: but (4) assumes that we have already been able to define this limit fractal, which may not be the case. From the point of view of physics, which is our final aim here, the problem is still more radical: the strict limit $\varepsilon = 0$ may have no physical sense at all.

The second relation expresses the fractal character of the function. Indeed f is assumed to be a finite fractal function, whose fractal behaviour is seen in the divergence of its derivative as ε approaches zero. The form of (10) expresses the facts that this derivative is defined for any $\varepsilon \neq 0$, but undefined for $\varepsilon = 0$, that there is a fractal-standard transition around $\varepsilon = \lambda$, and that the fractal dimension of the function f is $D = 1 + \delta$ when $\varepsilon = \Delta f$ and $D = \frac{1}{1+\delta}$ when $\varepsilon = \Delta x$. The quantity $\xi(x, \varepsilon)$ is a 'renormalized' derivative for the function f and is itself a new finite fractal function. For $\varepsilon \gg \lambda, \xi(\frac{\lambda}{\varepsilon})^\delta$ vanishes and df/dx reduces to a standard function $\varphi(x)$. When $\varphi = 0$, we get a transition to macroscopic flatness.

There is an interesting theorem about ξ, the complete demonstration of which will be given elsewhere (Nottale and Lachaud 1991): *the fractal dimension of the renormalized derivative of any fractal function is 2.* This means that:

$$\frac{d\xi}{dx}(x; \varepsilon) = \zeta(x; \varepsilon)\left(\frac{\lambda}{\varepsilon}\right) \tag{11}$$

with ζ finite, and this will hold for any of the following derivatives. Another property of ξ derives from a reintegration of (10). From the hypothesis that f is finite, we conclude that $(\frac{\lambda}{\varepsilon})^\delta \int \xi(x; \varepsilon)dx$ is itself finite, so that the integral of ξ is vanishing with ε as ε^δ, contrarily to the integral of $|\xi|$ which is finite and non null. A more extensive account of the various properties and applications of these fractal functions (Fourier transform, generalization to several variables, existence of fractal functions quasi-identical to their renormalized derivative, fractal solutions to differential equations, ...) will be the subject of a forthcoming paper (Nottale and Lachaud 1991).

Let us finally come back to our initial motivation, that is the definition of a fractal space. The possibility to replace the non differentiable fractal F by an equivalence class of families of differentiable functions can be now applied to geometry. For any given resolution ε, we are now back in the frame of differentiable geometry, so that the approximation of the fractal space may be taken Riemannian. We finally define a fractal space as the equivalence class of a family of Riemann spaces, characterized by a family of Riemann tensors $R_{ijkl}(\varepsilon)$. The fractal limit itself is undefined, being non differentiable and then non Riemannian (its curvature will be everywhere infinite). For each resolution ε, a coordinate system is given (of which ε is the 'state of resolution' or 'state of scale'), $x^i(\varepsilon)$, in which metric potentials $g_{\mu\nu}(\varepsilon)$ are defined, and from which the Riemann tensor coefficients may be computed by the usual expressions. The hereabove relations obtained for fractal functions give some hints about the way the various spaces at different ε will be related. A detailed account of the properties of such a geometry exceeds the frame of this contribution and will be published elsewhere.

4. The Fractal Dimension of Quantum Particle Paths

The question of the fractal dimension of a quantum mechanical path in the non relativistic case has been considered by Abbott and Wise (1981), Campesino-Romeo et al. (1982), Allen (1983) and others (see Nottale 1989). They found that quantum mechanical paths are characterized by a universal value of the fractal dimension $D = 2$, with a transition from non-fractal (classical, $D = T = 1$) to fractal (quantum, $T = 1$, $D = 2$) occurring around the de Broglie length,

$\lambda_{dB} = \hbar/p$. Let us recall briefly the argument to show how this result is a direct consequence of the position-momentum Heisenberg relation.

The total length travelled in the average during a set of experiments where the successive positions of a particle are measured with a resolution Δx is given by:

$$L \propto < |v| > \propto < |p| > \tag{12}$$

in the non relativistic case. The classical case $< |p| > \approx p_0 \gg \Delta p$, i.e. $\Delta x \gg \lambda_{dB}$ (since $\Delta x \Delta p \approx \hbar$) yields as expected a length independent from Δx. On the contrary the quantum case $\Delta p < p_0$ (i.e. $\Delta x < \lambda_{dB}$) yields $< |p| > \approx \Delta p$, so that the length diverges as:

$$L \propto \frac{\lambda}{\Delta x} \tag{13}$$

which corresponds to $D = 2$, since $T = 1$, see (2).

The relativistic case is more difficult to deal with. A first analysis may lead to the conclusion that, because of the limitation $v \leq c$, the length will become bounded again for $\Delta x \lesssim \lambda_c = \hbar/mc$, so yielding back $D = 1$ (Cannata and Ferrari 1988). However one should also account for the radically new physical behaviour which takes place in the quantum relativistic domain, i.e. virtual particle-antiparticle pair creation and annihilation. I have proposed to reinterpret the virtual pairs as a manifestation of the fact that the fractal trajectory is now allowed to run backward in time for time intervals $\Delta t \lesssim \tau_{dB} = \hbar/E_0$ (Nottale 1989). This is based on the Feynman-Stueckelberg-Wheeler interpretation of antiparticles as particles running backward in time. However here we consider electron-positron virtual pairs to be part of the nature of the electron itself, in agreement with the QED expression for the electron self-energy, which contains all successive Feynman graphs (Landau and Lifshitz 1972), but they are now interpreted as a manifestation of a unique self-avoiding trajectory in space-time (Nottale 1989). Indeed consider that for a time interval Δt, an energy fluctuation $\Delta E \approx \hbar/\Delta t$ may give rise to the creation of n e^+e^- pairs, with $\Delta E \approx 2nmc^2$. With the hereabove reinterpretation, the total time elapsed on the fractal trajectory is now given by the sum of the proper times of the $(2n + 1)$ particles, i.e.:

$$T = (2n + 1)t_0 = (E + \Delta E)\frac{t_0}{mc^2} = \left(1 + \frac{\tau_{dB}}{\Delta t}\right)t_0 \tag{14}$$

So we get a temporal non-fractal $D = 1$ to fractal $D = 2$ transition around the de Broglie time τ_{dB} (Nottale 1989). The argument holds also for the total distance travelled, $L = (2n+1)L_0 = (1 + c\tau_{dB}/\Delta x)ct_0$, which demonstrates that the spatial fractal dimension remains $D = 2$ in the relativistic case. Hence the transition from quantum non relativistic to quantum relativistic may be accounted for by a purely temporal transition. The length increase is now compensated by a time increase, which results in the relativistic bound $v \leq c$. Finally we get a Lorentz covariant scheme with a unique spatio-temporal transition from $D = 1$ to $D = 2$ around $\lambda^\mu = \hbar/p^\mu$, $\mu = 0$ to 3.

This result may also be obtained by the inverse argument: starting from the hypothesis of underlying fractal structures, the Heisenberg relations may indeed be found as a direct consequence of these structures, provided the fractal dimension is $D = 2$ and the transition occurs around de Broglie's wavelength (Nottale 1989). The argument may be summarized as follows: we start from an action $S(x^\mu)$ written in terms of 4 space-time coordinates. We now assume that the classical coordinate x^μ is classical only for resolution larger than some transition value λ^μ to be determined, but that it owns fractal structure at smaller resolution. This is described by the setting of some fractal coordinate X^μ of which x^μ is a fractal function: $x^\mu = x^\mu(X^\mu; \Delta x^\mu)$. We may write:

$$p_\mu = -\frac{\partial S}{\partial x^\mu} = -\frac{\partial S}{\partial X^\mu}\,\frac{\partial X^\mu}{\partial x^\mu} \tag{15}$$

But X^μ is multivariate in terms of x^μ, so that the last term is relevant of the fractal derivative definition (3). Taking the quadratic mean of the momentum, we get (here the Einstein notation is not applied):

$$\Delta p_\mu (\Delta x^\mu)^\delta = \left[(\lambda^\mu)^\delta p_\mu^0\right] \, \Delta x \tag{16}$$

Indeed the classical conservation of 4-momentum is known to be a consequence of the homogeneity of space-time. The postulated fractal structure of space-time at the quantum level breaks this uniformity, and this is translated in the Heisenberg inequalities. But, as remarked in Sect. 3, homogeneity is recovered locally as Δx^μ approaches 0 (strictly at the non-standard infinitesimal level), so that $-\partial S/\partial X^\mu$ is a conserved quantity which we shall identify with the classical momentum p_μ^0. The quantity $\Delta\xi$ is dimensionless. Assuming it is always larger than $1/4\pi$, setting $\delta = 1$, i.e. $D = 2$, and identifying λ^μ with the de Broglie length and time finally yields the Heisenberg relations. But the interest of (16) is that it is an equality instead of an inequality, and that it may become the basis of a generalisation of Heisenberg's relations (Nottale 1991).

One of the interesting consequences of this interpretation is that it precisely accounts for the Zitterbewegung, this oscillatory motion of the center of mass of an electron resulting from the Dirac equation. This effect is known to be the result of interactions between the negative energy and positive energy solutions of the Dirac equation. Though it indeed disappears if one keeps only the positive energy solutions to describe an electron, this does not yield a satisfactory solution to the problem, since such a positive energy electron would be completely delocalized (Bjorken and Drell 1964). Conversely one may set the localization assumed (or measured) for the electron, $\Delta x = c\Delta t = \hbar c/E$, and then derive the relative rate of positive and negative energy solutions. One gets:

$$\frac{P_-}{P_+} \approx \left[\frac{pc}{E + mc^2}\right]^2 = \frac{E - mc^2}{E + mc^2} \tag{17}$$

Now in the fractal model, for each classical time interval we have $2n + 1$ segments, $n + 1$ running forward and n running backward, so that with $E = (2n + 1)mc^2$, one gets $P_-/P_+ = n/(n+1) = (E - mc^2)/(E + mc^2)$, i.e. *exactly the QED result.* This opens the possibility to build fractal localized one particle solutions to the Dirac equation (Nottale 1991).

5. Geodesical Interpretation of the Wave-Particle Duality

Let us define the de Broglie length and time (λ, τ) as geometrical structures of the fractal trajectory of a 'particle': they are identified with the fractal/non fractal transition. Then the various classical quantities may be expressed in term of these two geometrical quantities:

$$E = \frac{\hbar}{\tau} \; ; \; p = \frac{\hbar}{\lambda} \; ; \; v = \frac{c^2 \tau}{\lambda} \; ; \; m = \frac{\hbar}{c}\sqrt{\frac{1}{c^2\tau^2} - \frac{1}{\lambda^2}} \tag{18}$$

This means that we do not have to endow the point particle with mass, energy, momentum or velocity, but instead that these properties may be reduced to geometrical structures of the particle trajectory. As a consequence the energy- momentum tensor (Weinberg 1972), and thus the right hand term of Einstein equations, may be written under a completely geometrical form in terms of the de Broglie periods and of the Planck length λ_p (with $X_i(\sigma_i)$ being four fractal functions defining the trajectory of particle i in terms of a curvilinear fractal proper time σ_i):

$$\frac{G}{c^2} T^{\mu\nu} = \sum_i \int c\tau_i \, \frac{\lambda_p}{\lambda_i^\mu} \, \frac{\lambda_p}{\lambda_i^\nu} \, \delta^4 \left[x - X_i(\sigma_i) \right] \, \frac{dt}{d\sigma_i} \, d\sigma_i \tag{19}$$

We have also demonstrated that the quantum spin itself may be obtained as a proper angular momentum of fractal trajectories having precisely fractal dimension 2 (Nottale 1989). This result is particularly interesting since in the present quantum theory the spin is considered as a purely quantum and internal quantity without any classical analog (at least as concerns the particle aspect). This is due to the fact that a particle like the electron is assumed to be totally point-like, and that clearly a non extended object can have no classical proper angular momentum. This is where the non-classical character and the infinities of fractals intervenes: a point 'particle' following a fractal trajectory may own a spin even though its radius is zero, because its rotational velocity is infinite. This may be understood by the following simplified argument.

Assume that the fractal trajectory is defined from a generator made of p segments of length $1/q$. Consider the approximation F_{n+1} of the trajectory, having resolution q^{-n}. We may now compute the average angular momentum of the particle following F_{n+1} with respect to the trajectory F_n. Assume that we get $M_n = \sum mr^2\dot{\varphi}$. When going to the next stage of the fractalization, if one wants to compute now the angular momentum of F_{n+2} with respect to F_{n+1}, all the distances will be divided by q, while the angular velocity will be multiplied by p, yielding: $M_{n+1} = M_n(pq^{-2})$. By iteration we find that only the case $pq^{-2} = 1$ may yield a finite angular momentum at the fractal limit. But $D = \log p/\log q$, so that while for $D < 2$ and $D > 2$ the proper angular momenta of fractal curves are respectively 0 and infinite (i.e. cannot be defined), it becomes finite for the strict value $D = 2$. It may be shown that it also owns all the properties of an *internal* quantum number (Nottale 1989). The interesting point here is that $D = 2$ was already obtained from Heisenberg's relations alone, which do not assume a spin, so that we thus get an independent check of this result and of its physical significance.

Let us now develop the geodesical interpretation of the wave-particle duality, and show that it may help understand the basic quantum mechanical paradoxes. A fractal space-time is characterized by an infinity of obstacles at all scales and by returns and eddies also at all scales and for all 4 coordinates. As a consequence one expects that any two space-time events will be connected, not by one geodesic only, but by an infinity of geodesics, so that the properties of physical objects on a fractal space-time will result in a mixing of individual properties (one particular geodesic) and collective properties (those of families of geodesics), in agreement with the wave-particle duality. Individual geodesics shall be also partly endowed with wave properties, because of their fractal character from which the de Broglie length and time may be defined, as recalled in previous sections, but the full wave function is expected to characterize the complete family of geodesics.

While one may admit, without any logical inconsistency, that a particle follows one particular geodesic (as indicated by individual *measurements* of localized particles), one must admit at the same time that all geodesics are equiprobable (as expected in a geometrical space-time theory), so that any *prediction* can be only of statistical nature, since applying to a family of geodesics. So in such a frame we get the possibility, at least in principle, to have both a space-time which would be stricly determined by its material and energetic content, and a definitive loss of determinism concerning particle trajectories. But here the statistical behaviour would not be the cornerstone on which physical laws are based, but instead a consequence of the fractal structure of space-time.

One result supporting this interpretation is that indeed beams of geodesics may have a wave nature described by a Schrödinger-like equation in a Riemannian space-time, even in the geometric optics approximation (Nottale 1988b, 1989). A light beam in general relativity is described at the geometric optics approximation by a congruence of null geodesics, the equations of which have been written by Sachs (1961). The cross sectional area A of the light beam may be subjected on its way to three infinitesimal deformations, expansion $\theta \cdot d\omega = d\sqrt{A}/\sqrt{A}$, rotation $\Omega \cdot d\omega = dW$ and shear. Considering only the shearless case, when setting $\psi = \sqrt{A} \cdot e^{iW}$ the beam propagation equation writes, in term of an affine parameter ω and of Ricci tensor R_{ij} and wave vector k^i (Nottale 1989):

$$\frac{d^2\psi}{d\omega^2} + \left[\frac{1}{2} R_{ij} k^i k^i\right] \psi = 0 \tag{20}$$

which has exactly the form of the one-dimensional Schrödinger equation. So a family of geodesics, even in the geometric optics approximation, possesses the equivalent of a quantum phase, interpreted as the beam rotation and of a probability of presence, identified to the beam cross sectional area.

Another possible approach consists in remarking that the effect of the fractal structures of space-time on the trajectory may be identified with some diffusion process. We would thus obtain a stochastic version of the theory, but into which the physical cause of the underlying quasi-Brownian motion would be directly understood. Such a connection to stochastic quantum mechanics has the advantage to demonstrate in a straighforward way the ability of the theory to yield quantum mechanics at some approximation level, since stochastic quantum mechanics is already known to be relevant for the Schrödinger equation and to yield the same predictions as quantum mechanics (Nelson 1966, 1985). But it is only in the completely consistent approach using the geometrical tools of fractal space-time, geodesics and geodesic deviation equations (à theory which still remains to be developed), that one may in our view expect possible new predictions.

Let us consider now some of the basic quantum experiments and show how their paradoxical properties are enlighted by the fractal geodesics interpretation.

Young's holes experiment. This is certainly one of the most fundamental quantum experiments, which shows both the wave-particle nature of quantum objects and the phenomenon of complementarity. Let us first comment about the so-called Copenhagen interpretation of this experiment and of quantum mechanics. The fact that, if one wants to observe the interference pattern, it becomes impossible to know by which hole did the particle pass, has been interpreted as indicating that there was no value of truth (or of reality) into asserting that it indeed passes through one or the other hole. By extension this leads to the idea that no reality should be attributed to the particle between the measurements, and finally that the measurement results are determined by the measurement apparatus and just at the time of measurement (rather than revealing a reality which was preexisting to the measurement). This interpretation assumes that there can be no value of truth attributed to something which is either non measurable or non predictible. Concerning at least the domain of predictions, which comes under mathematical physics, one may remark that, as such, it should be dependent on Gödel's theorem (1931). We recall that this theorem implies that in any consistent system based on a set of axioms which contains number theory, there exist non demonstrable (true) sentences. Its main consequence is that it has definitively demonstrated the logical difference between truth and provability. We suggest that some of the paradoxes of quantum mechanics reveal the first intrusion of this limitation into mathematical physics. Applied to the Young's hole problem, this means that our inability of predicting or retrodicting by which hole did the particle pass does not imply than it is not true that it passed through one of the holes. Similar conclusions seem to be reached in the logical reformulation of quantum mechanics by Omnes (1988).

The understanding of Young's holes in terms of geodesics of a fractal space-time goes as follows. For each configuration (one hole open, two holes open, detector behind one hole, see Feynman et al. 1965) one should consider the whole infinite set of geodesics which connect the source, the possible intermediate detector and the final point of detection of the particle on the screen. The final probability distribution of impacts on the screen will be given by the relative ratios of the (infinite) number of geodesics. So it is clear that any detection of the particle before the screen reduces the set of geodesics to those which pass through the detection point and so destroys the interference pattern. The fact that the number of geodesics is always infinite also allows to understand how, when two holes are open instead of one, there is no particle detected at points of destructive interference, while an increased number of geodesics is expected: indeed we may have both an increase of the number of geodesics from the one hole to the two hole experiment, and a relative ratio of the number of geodesics which is itself infinite between the points of constructive and destructive interference. In fact this virtual set of geodesics fills space like a fluid, so that it becomes easy to endow it with wave properties. When an experiment is actually run, either the number of particles is large (sometimes even undefined), and the structures of the geodesics are immediately visualized, or particles arrive at the detector one by one, and the interference pattern will be little by little achieved in a probabilistic way.

EPR Paradox. The EPR paradox, with its explicit violation of Bell's inequalities (Aspect et al. 1982), plays a central role in our understanding of quantum mechanics. Consider the anticorrelated emission along two separated beams of two particles in a mixed state of some quantum number s which may take two values, say +1 and −1. The state in which the two particles are emitted is such that $\{P_1(+1) = 1/2, P_1(-1) = 1/2\}$, $\{P_2(+1) = 1/2, P_2(-1) = 1/2\}$ and $s_1 + s_2 = 0$. There is no way to know in the absence of measurement whether particle 1 (or 2) has $s = -1$ or $+1$, but any measurement of s_1 alone allows to immediately infer that $s_2 = -s_1$. In the present interpretation of quantum mechanics, the mixing of the state is a property which is attributed to the particle itself. In these conditions, nothing more than $[P_1(+1) = 1/2, P_1(-1) = 1/2]$ can be said of the particle prior to a measurement of s, so that it becomes clear that the value of s given by the measurement does not prexist to it. The determination of s_2 being simultaneous with that of s_1, we thus get the usual paradoxical result, either of strong non locality, or of propagation of information faster than the velocity of light. Let us propose another interpretation. As recalled hereabove, in the geodesical interpretation, the properties of quantum systems will consist in a mixing of individual properties, those of individual geodesics and of collective properties, those of the families of equivalent geodesics from which predictions can be drawn. Let us then assume that a mixed state is made of a beam containing a fractal distribution of geodesics, each of which is defined by a well determined value of s. Assuming that the distribution is fractal means that, in whatever domain either of space or of phase space or of any other physical quantities considered to whatever resolution, there will always be found a mixing of an infinite number of +1 geodesics and −1 geodesics. This is a (nondifferentiable) model in which there exists no hidden parameter which could allow to tell the value of s prior to a measurement, but in which we may also admit that one particular geodesic has actually been followed by the particle (note that differentiability is assumed in the setting of Bell's inequalities). In this interpretation a measurement of the kind described above does not determine the measured quantity, but reveals its preexisting value, even though we should admit our definitive inability to predict it. It should be remarked that quantum non locality is kept in this interpretation, as a property of the family of geodesics.

Indiscernability. The indiscernability of equivalent particles is a straighforward consequence of the identification of 'particles' to the geometric structures of fractal trajectories themselves.

6. Possible Consequences for Gravitation

Here we come to investigating the connection of our approach with the theory of general relativity: it is from such a study that one may expect to derive possible new astrophysical and/or cosmological consequences. One important question which was left untackled by general relativity is the way the sources of gravitation act. The strong equivalence principle postulates the total identity between the active gravitational mass, the passive gravitational mass and the inertial mass. While the equivalence between the last two is completely understood from the structure of the theory itself (a reference system in free fall is locally inertial, so that there is in fact no need for the definition of a passive gravitational mass: the inertial mass immediately plays this role), nothing is known about how masses (and more generally energy-momentum) curves space-time. Anyway the strong principle is now experimentally verified to a high degree of accuracy.

Another related and important question which may still not be considered as closed is Mach's principle and the hope that it brought to describe all motions as relative and understand the amplitude of inertial forces (i.e. eventually the value of G) from the distribution of masses in the Universe, a hope which was not achieved by general relativity. To illustrate this frustration, consider a rotating body, flattened in the direction of its axis of rotation. We know that this flattening comes from the centrifugal force of inertial origin. This force can be understood only if we are able to tell how is defined the inertial system relative to which the body is rotating. Mach's and Einstein's analysis rejects Newton's answer that it would be turning relative to absolute space, since such an absolute space can not be physically characterized and is at variance with the principle of relativity. Mach's answer (awaiting a detailed theory) is that inertia results from masses distributed at infinity. This means that if one wants to transform Mach's principle into a self-consistent theory, strictly such a theory should not allow the existence of mass distributions (i.e. of Universes) devoid of masses at infinity. Einstein's answer is that the whole distribution of masses in the Universe determines the curved space-time, so that the body is turning with respect to some reference system which is in 'free fall' in the gravitational field so determined. While this answer is a definite, satisfactory (and experimentally verified, Weinberg 1972) explanation if the rotating body may be considered a test object in a Universe containing other masses, general relativity nevertheless does not solve the problem *in principle*, since it allows solutions of the field equations empty everywhere except for a single mass, possibly rotating. The active (and then inertial) mass in Schwarzschild's solution comes out from the equations as an integration constant; its existence may be related to no mass distribution, since the metric is Minkowskian at infinity; still worse, the rotation of the active gravitational mass in a Kerr solution is defined in a quasi-absolute way, space-time exterior to it being empty: there is no objective reference with respect to which a Kerr mass is rotating (and then is subjected to centrifugal force), an unphysical situation against which general relativity, in its premises, was expected to fight (Einstein 1916).

Is the fractal space-time approach able to bring new insights about these problems? Several arguments seem to open such a hope. The ability it offers to structure the 'internal' part of objects in a geometric way leads to the idea that gravitation, i.e. the manifestation of curvature, might be described as a residual effect of these structures for macroscopic distances and masses. Indeed fractalization, as recalled hereabove, is characterized by an 'hypercurvature', infinite everywhere but flipping in a fractal way from one sign to the other. The result of averaging this hypercurvature at large scale (i.e. $+\infty - \infty$) may yield flatness, but more generally any finite curvature. To be more specific, the obtention of a general relativistic space-time as a limit of a fractal space-time is a direct consequence of our description of a fractal space-time as a family of Riemannian space-times, $R_{ijkl}(\varepsilon) \rightarrow R_{ijkl}$ independent of ε when $\varepsilon \rightarrow \infty$. The form obtained in (19) for the right-hand side of Einstein equations is also interesting in this connection: it is symmetrical in G and $\hbar$, and may be read as totally geometrical in nature.

Let us now develop a general argument which demonstrates that the present structure of physical theories, of general relativity as well as of quantum mechanics, needs revision, and this maybe not only at the Planck scale. The wave nature of any physical system is at present assumed to be universal. However the wave properties of a system keep a physical meaning only if one may, at least in a 'Gedankenexperiment', measure them (i.e. by a diffraction or interference experiment). But when the total mass m becomes larger than about the Planck mass $m_p = (\hbar c/G)^{1/2} \approx 2 \cdot 10^{-5}$g, its Compton length becomes smaller than its Schwarzschild radius $r_s = 2Gm/c^2$, thus becoming unmeasurable, not only for technological limitation, but mainly for a profound physical limitation, since it enters into a black hole horizon (Nottale 1989). Indeed let us attempt at describing more precisely such an experiment: in order to evidence an interference pattern in e.g. a Young holes measurement, the hole width should be smaller than or at most of the order of the wavelength of the object which we want to make interfere. So it is first clear that we can not get interferences with objects which would be much larger than their de Broglie wavelength. To deal with this problem, and accounting for the fact that we are interested here only in the mass dependence of the wave behaviour, independently from the nature of the object, let us assume that we have reduced the mass into a point. Even in this case, when the mass becomes larger than the Planck mass, it is anyway surrounded by its Schwarzschild sphere: if the experiment was to be actually made, the black hole horizon being larger than the holes would break them, thus preventing us from pursuing the experiment. Strictly it is the Compton length which becomes smaller than the Schwarzschild radius, and one may argue that with a low enough velocity one would get a large enough de Broglie length. There are (at least) two answers to this: first in such a (Gedanken) experiment, the resolution of the measurement apparatus should be of the order of the Planck length; then the Heisenberg relation implies $p \gtrsim \hbar/\lambda_p \simeq m_p c$, and since we deal with a mass of the order of the Planck mass, its velocity is close to that of light; second the de Broglie length $\hbar/p$ derives from the de Broglie time in rest frame $\hbar/mc^2 = \lambda_c/c$ through a Lorentz transform. Therefore, if the Compton length loses its physical meaning, so will the de Broglie length.

Conversely, for $m < m_p$ the Schwarzschild radius becomes smaller than the Compton wavelength. But for smaller resolution than the Compton length, we enter into the quantum relativistic domain inside which the concept of position itself loses its physical meaning (Landau and Lifshitz 1972), because of particle-antiparticle creation. This means that one cannot localize a particle with a resolution better than its Compton length. However a black hole state can be reached only provided the whole mass m is confined into its Schwarzschild radius. So it appears that the concept of Schwarzschild radius loses its physical meaning for $m < m_p$.

In Nottale (1989), it was attempted to connect the fractal approach to general relativity by the following way. Let us consider a simple model of fractal metric:

$$ds^2 = \xi^2(t, \Delta t)\left(1 - \frac{\tau}{\Delta t}\right)^2 c^2 dt^2 - dl^2 \tag{21}$$

in which ξ is some finite fractal function (with $\xi = 1$ for $\Delta t > \tau$), $\tau = \hbar/mc^2$ is the de Broglie time of a system of inertial mass m and dl^2 is a fractal spatial element which will not be more detailed here. This form of g_{00} has been chosen for the following reasons: it yields fractal dimension 2 for $\Delta t \ll \tau$, the Minkowski limit $g_{00} = 1$ for $\Delta t \to \infty$, and a singularity $g_{00} = 0$ for $\Delta t = \tau$, thus accounting for creation/annihilation of particles (in analogy to the cosmological primeval singularity). Consider the domain $\Delta t \gg \tau$, g_{00} may be expanded as $g_{00} = 1 - 2\tau/\Delta t$. Let us now identify $\Delta t \equiv r/c$ to the lifetime of an exchanged virtual boson (in analogy with the simple well-known calculation which yields Coulomb law: $F \simeq \delta p/\delta t \simeq (\hbar/r)/(r/c) \simeq \hbar c/r^2 \simeq e^2/r^2$). One gets:

$$g_{00} = 1 - \frac{2\hbar}{mcr} \tag{22}$$

This result holds for one particle of mass m, but also for a complex system of total mass $\sum m$, thanks to the universality of the de Broglie wave nature of any physical system. But the hereabove arguments suggest that both (22) $g_{00} = 1 - 2\hbar/mcr$ and the Schwarzschild potential $g_{00} = 1 - 2Gm/c^2 r$ become unphysical, respectively for $m > m_p$ and $m < m_p$. The remarkable result here is that *they are precisely equal for* $m = m_p$.

We thus proposed (Nottale 1989) that there exists some fundamental microscopic / macroscopic transition around the Planck mass, which would hold not only at the Planck length scale, but also at macroscopic length scale and would connect the quantum behaviour to the gravitational (general relativistic) one. This would be e.g. achieved by the following phenomenological 'potential':

$$g_{00} = 1 - 2\left(\frac{m}{m_p} + \frac{m_p}{m}\right)\frac{\lambda_p}{r} \tag{23}$$

However the physical meaning of (23) is by now still unclear. If in the 'gravitational' regime it indeed plays the role of a metric potential which may be connected to the Newtonian potential, its interpretation as a potential in the microscopic regime does not seem to be acceptable.

The way to clarify such questions is now to turn to experiment. As a possible observable consequence, the Newton law should break down between two dust particles both having mass smaller than the Planck mass (Nottale 1989). This could be checked in a space experiment: for example two 2.10^{-5}g dust grains of density $d = 18$ have a radius of 0.064mm. Their expected Newtonian free fall time for an initial relative distance 0.24mm would be $\approx$ 10mn. We expect the falling time to become smaller than the Newtonian time for $m \lesssim m_p$. Another possibility would be to build a micro 'solar system' with a proof mass $m \ll m_p$ orbiting around a Planck mass, and to follow its temporal evolution. Such an experiment would be at the limit of the present possibilities, needing a high control of vessel gravity gradients and of electric charges, since the electric force would be equal to the $m_p \times m_p$ gravitational one for only $\alpha^{-1} \approx 12 \times 12$ elementary electric charges.

The hereabove calculation is a very rough one including only temporal terms in the metrics, in which the mass considered was assumed to remain a point mass. We intend to attempt its spatio-temporal generalization to the extended case. A possible expectation of the result would be the generalization of the Planck mass transition to a critical density transition (Nottale 1991). Then astrophysical and cosmological implications, concerning e.g. the formation of structures or the dark matter problem, are expected to be derived in such a constrained way that the hereabove suggestion of a transition between a quantum and a gravitational régime could be either shown to be self-consistent with present knowledge, or easily falsifiable.

Acknowledgements

I gratefully acknowledge Myriam Marouard's invaluable help in the TEX processing of this contribution and Yves Lachaud for stilumating discussions during its preparation.

References

Abbott, L.F., Wise, M.B. (1981): *Amer. J. Phys.* **49**, 37.
Allen, A.D. (1983): *Speculations Sc. Techn.* **6**, 165.
Aspect, A., Dalibar, J., Roger, G. (1982): *Phys. Rev. Letters* **49**, 1804.
Bjorken, J.D., Drell, S.D. (1964): *Relativistic Quantum Mechanics* (McGraw-Hill, New York).
Campesino–Romeo, E., D'Olivo, J.C., Socolovsky, M.(1982): *Phys. Letters* **89A**, 321.

Cannata, F., Ferrari, L. (1988): *Amer. J. Phys.* **56**, 721.
Einstein, A. (1907): *Jahrb. Rad. Elektr.* **4**, 411;
Einstein, A. (1916): *Ann. Physik* **49**, 769.
Einstein, A. (1948): *Letter to Pauli*, in *Quanta* (Seuil/CNRS, Paris), p. 249.
Feynman, R., Hibbs, A. (1965): *Quantum Mechanics and Path Integrals* (McGraw–Hill, New York).
Feynman, R., Leighton, R.B., Sands, M. (1965): *The Feynman Lectures on Physics III* (Addison-Wesley, Reading).
Fuller, F.W., Wheeler, J.A. (1962): *Phys. Rev.* **128**, 919.
Gödel, K. (1931): in *Le Théorème de Gödel* (Seuil, Paris).
Hawking, S. (1978): *Nucl. Phys.* **B 144**, 349.
Landau, L., Lifshitz, E. (1972) *Relativistic Quantum Theory* (Mir, Moscow).
Le Bellac, M. (1988): *Des Phénomènes Critiques aux Champs de Jauge* (InterEdition/CNRS, Paris).
Mandelbrot, B. (1982a): *The Fractal Geometry of Nature* (Freeman, San Francisco), p. 331.
Mandelbrot, B. (1982b): *The Fractal Geometry of Nature* (Freeman, San Francisco), p. 365.
Nelson, E. (1966): *Phys. Rev.* **150**, 1079.
Nelson, E. (1977): *Bull. Amer. Math. Soc.* **83**, 1165.
Nelson, E. (1985): *Quantum Fluctuations* (Princeton Univ. Press, Princeton).
Nottale, L. (1988a): *C.R. Acad. Sc. Paris* **306**, 341.
Nottale, L. (1988b): *Ann. Phys. Fr.* **13**, 223.
Nottale, L. (1989): *Int. J. Mod. Phys.* **A4**, 5047.
Nottale, L. (1991): submitted.
Nottale, L., Lachaud, Y. (1991): in preparation.
Nottale, L., Schneider, J. (1984): *J. Math. Phys.* **25**, 1296.
Omnes, R. (1988): *J. Stat. Phys.* **53**, 893, 933, 957.
Robinson, A. (1961) *Proc. Roy. Acad. Sc. Amsterdam* **A 64**, 432.
Sachs, P.K. (1961): *Proc. Roy. Soc. London* **A 264**, 309.
Sagdeev, R.Z., Usikov, D.A., Zaslovski, G.M. (1998): *Non–Linear Physics* (Harwood, New York).
Weinberg, S. (1972): *Gravitation and Cosmology* (Wiley, New York).
Wilson, K.G. (1979): *Scientific American* **241**, 140.

The Real Stuff*

André Heck

Observatoire Astronomique, 11 rue de l'Université, F-67000 Strasbourg, France

> [According to NASA], scientists have looked at only 10 percent of the data sent back to Earth in the past 20 years, and have analyzed only 1 percent.
>
> *(Newsweek, 7 May 1990)*
>
> The Hubble Space Telescope is expected to send down yearly to Earth the equivalent of about 500 Gbyte of data.
>
> *(see e.g. Russo et al. 1986)*

Abstract: The present situation in the field of astronomical databases and archives is discussed. Shortcomings and possible future trends are mentioned, as well as some considerations on a future ideal environment for direct home applications on real and remotely located astronomical data.

1. Introduction

Astronomical theory and methodology would remain sheer rhetoric without the observational data with which they can be confronted and to which they can be applied. The present chapter aims at giving an overview of the present situation in the field of astronomical databases and archives.

A few shortcomings will be mentioned together with possible solutions and future trends that should ease their resulting constraints. A touch of dream will conclude this paper.

Databases are now linked to multiple aspects of our activities: from the observing proposal elaboration, submission and evaluation, to the reduction of new data

---

* With apologies to Tom Wolfe.

with or without integration of data obtained previously or with different instrumentations, and to the finalization of papers implying at each step the collection of a maximum of isolated data of various types and from different sources, without forgetting the cross-checks with what has already been published on the objects studied or on similar ones.

All this has to be put in parallel with the dramatic revolution in scientific communications, expressed recently by the explosion of electronic mail and networks, as well as with the desktop publishing facilities, which were not at all in practice only a few years ago. This allows quick, efficient and high-quality dialogues, and consequently publishing, with remotely located collaborators.

2. Databases, Archives and Networks

It would be pretentious to intent to give here in a few lines an exhaustive list (not to speak about a description) of all astronomical databases and archives presently available to the professional astronomer. The situation is evolving quite rapidly in this field and the most recent and up-to-date picture of the situation is given in a whole book compiled by Albrecht and Egret (1991).

The general understanding for an *archive* is in a context of a set of data obtained through a given experiment, be they on a classical support as photographic plates or digitalized on magnetic tapes or other supports. These data might not be anymore quite *raw*, in the sense that they might have undergone some amount of treatment or *processing*.

Typically archives would be for example:

- the set of ultraviolet spectra collected by the *International Ultraviolet Explorer (IUE)* (see Wamsteker 1991);
- the European Southern Observatory (ESO) archiving project which has started systematically with the New Technology Telescope (NTT) operations (Ochsenbein 1991);
- the National Radio Astronomical Observatory (NRAO) archives from its various facilities (Wells 1991).

These are just three examples relative to different wavelength ranges (respectively ultraviolet, visible, and radio), but the list could be much longer and include the archiving aspects of other experiments such as COBE (White and Mather 1991), EXOSAT (White and Giommi 1991), GRO (den Helder 1991), HIPPARCOS (Turon et al. 1991), HST (Schreier et al. 1991), IRAS (Walker 1991), ROSAT (Zimmermann and Harris 1991), and so on.

The term *database* could be generic and include archives as well. A stricter meaning however would link it to a context of scientific data extracted or derived from observational material after reduction and/or analysis, although a database can also be made of non-observational data. Contrary to a *databank*, a database would also generally contain some software to retrieve the data and possibly work on them.

A typical database is the pionneering SIMBAD hosted by the Astronomical Data Centre (CDS) at Strasbourg Observatory (Egret et al. 1991).

Some experiments have got a database built around their archive and it is sometimes difficult to distinguish the two aspects. This is particularly the case for the EXOSAT Database (White and Giommi 1991) and the NASA/IPAC Database (Helou et al. 1991).

Pionneering networks such as ESA's ESIS (Albrecht 1991) and NASA's ADS (Weiss and Good 1991) aim at interconnecting and making easily available the most relevant archives and databases to the astronomical community. See Murtagh (1988; 1991) for reviews on current issues and status.

The line 'speed' (or better 'data flux capacity') should still be increased in order to accomodate quick transfers of huge amounts of data. The transfer of bi-dimensional images remains still prohibitive nowadays, but similar requirements in medical communications might have a stronger push on an advance in that respect than ours. Networks should absorb more and more databases and archives, with more and more diversified types of data (see Sect. 8).

3. Provisions for Databases and Archives

Here are a few recommendations that could be formulated for ground and space experiments in relation with the data they would collect (already partially formulated in Heck 1986):

- no project should be allowed to go ahead without proper provisions for an archive including data retrieval and dissemination at the end of a possible proprietary period;
- this possible proprietary period should be kept as reasonably short as possible;
- data availability should be as flexible as possible, as far as means of retrieval and supports are concerned; standard format (such as FITS – see Grøsbol 1991) should be highly recommended;
- data processing for plain astronomers (i.e. other than principal investigators, project designers, staff members, and so on) should be possible at the archive centers, possibly by remote logon and through decentralized pieces of software (see Sect. 5);
- adequate investigations and simulations should be run in advance to identify, as far as feasible, the appropriate pieces of data (or telemetry in case of space experiments) to be retained;
- the importance of a clean, complete and homogeneous log of observations cannot be stressed enough, as well as its easy integration in a network of databases;
- standard and calibration data should be made routinely available;
- special attention should be paid to a good representation (in the statistical sense) of all target categories observable with a specific experiment;

- before the possible termination of an experiment, plans should have been made for a complete reprocessing of all data collected with the latest and supposedly best version of the corresponding image processing software (IPS);
- the end of an experiment should not mean the turnoff of all activities related to the project as the corresponding database should be maintained with an appropriate service to the astronomical community;
- budget provisions should be made accordingly.

Last, but not least, one should not forget the invaluable service that a minimum quantity of software could provide to database or archive users. If each user cannot presently host in his/her local system an extensive IPS, a minimum processing capability and some mathematical – essentially statistical – methodological facilities should be implemented.

A statistical package such as the one available in MIDAS, the image processing system developed at ESO, has been derived from Murtagh and Heck's (1987) set of routines. It could be considered for a minimum exploitation of the data at hand. Other environments are described in Murtagh and Heck (1988).

4. Maintenance of Databases and Archives

The recommendations above imply a number of measures that go beyond mere archiving and dissemination of data: one could never stress enough the necessity of a good maintenance of databases and archives.

As indicated in the previous section, this leads to a number of reprocessings (whenever bugs have been detected or refinements can improve the IPS) and to the availability of a dedicated staff, even after the possible termination of the data collection period for an experiment with a limited lifetime.

Quality controls have to be carried out at various levels. In a database such as SIMBAD, the catalogues integrated undergo first a screening by a specialist or a team of specialists in the field. However even after integration of data checked in such a way, it is necessary to have a number of people controlling the databases regularly for purposes such as homogeneity, errors detected by users, and so on.

Artificial–intelligence, or more generally knowledge–based, procedures could be of a great help in this. Refer for instance to various papers in Heck and Murtagh (1989).

It is important that the database and archive managers stay in touch with their users. It might be appropriate to set up specific users committees and to run surveys regularly. Bulletin board services would be recommendable as well as mailbox systems to allow casual dialogues with users.

It might be the place to point out here the existence of a ***Working Group on Modern Astronomical Methodology*** and of a ***Technical Committee*** of the ***International Association for Pattern Recognition*** (TC13 – Astronomy and Astrophysics) (refer i.a. to Heck et al. 1985 and to Murtagh et al. 1987). The rôle of those groups is to set up links between interested people, to keep them posted on the advances in the corresponding methodologies, to suggest, support and help organizing meetings, workshops, schools, and so on.

5. Decentralized (Sub-)Databases and (Sub-)Archives

We have to face a paradox here. Indeed one could say that, because of the increasing facility to get connected to specific databases and archives, the idea of decentralized databases and archives (or subsets of these) could be completely discarded.

I believe that this is wrong for at least two main reasons. First, networks are not yet fully satisfactory as far as accessibility and line speed are concerned (not to speak of the so–called developing countries). Second, with the present impressive power of desktop computing facilities (in steady improvement), the difficulties for accomodating locally databases are becoming negligible. Refer for instance to what has been done with 'Einstein' data (Garcia et al. 1990; Plummer et al. 1991).

In parallel, the cost of producing massively CD-ROMs has gone down a lot and is decreasing rapidly. According to Garcia (1991), a thousand CD-ROMs would presently cost about US $ 5000. In view of this, one could only encourage all databases and archives managers to make available their babies on CD-ROMs together with an adequate piece of software. The progressive relative cheapness of such as system would also allow to keep the distributed (sub-)databases frequently updated.

According to Adorf (1991), there are now on the market high–performance image archival system consisting of rewritable CD–drive and software by *GigaSearch*: the disk stores up to 635 Mbytes with an access time of 58 msec. The corresponding catalogue can encompass 32,000 images. Additionally, each entry can be indexed with keywords allowing an efficient retrieval through logical expressions. This is definitely a way that need to be investigated.

Clear MOUs between data centres setting up collaborations have to be drawn with a minimum of rules guaranteeing a copyright protection and a good policy of acknowledgements in publications based on the various databases and archives, in such a way that everybody be happy – and essentially the funding institutions. This can also be considered as an aspect of the general security issue of databases.

6. Hierarchical Structure of Data

The situation in this respect is not yet quite satisfactory and significant steps should be made to improve it.

Let us take a simple example. If you are interested in a star in Orion, you surely want to know it belongs to this region, which is not obvious if you know the object only by its HD number or even more by a more obscure identification. Thus for a point–like object, you are keen to obtain data, such as those relative to the interstellar absorption, that correspond to an extended region it may belong to.

For a star cluster, you also wish to go upwards and downwards: from individual stellar data to the cluster ones, and from the cluster to the individual stars.

This is not easy to solve because another feature has to be taken into account: the resolution of data relative to extended objects. An infrared or an X-ray survey of the sky cannot always match the resolution obtained in other wavelengths. Therefore the correlation between databases cannot always be built up easily.

7. Integration of Wavelengths Ranges

Thanks to the use of balloons, rockets and spacecrafts, the part of the electromagnetic spectrum that has been more signicantly studied has been broadened. It is common now to carry out studies involving visible, UV, IR and X data, and to conduct joint observing campaigns involving ground-based instruments and spacecrafts working simultaneously in various ranges.

Therefore a coordination between archives and databases is also desired. A user must be able to mix data from various instrumentations in the most transparent way possible. Some fields of activity in astronomy, historically linked to specific wavelengths ranges, still remain too disconnected from the rest of the corporation.

8. Complementary Services

A network of interconnected databases and archives should not only provide scientific data, but also some of another nature, less scientific ones, but not least useful and practical, such as bibliographic and 'corporatists' ones.

One has already access to various kinds of bibliographic data on paper, like the invaluable *Astronomy & Astrophysics Abstracts (AAA)* (see e.g. the last four annual volumes by Burkhardt et al. 1990a,b,c,d), as well as bibliographic on-line databases (Watson 1991). These are accessible through authors and/or keywords. The bibliographic part of SIMBAD is object-oriented and thus of a complementary nature. All these facilities should be integrated.

Institutional data such as those provided by directories like ASpScROW (Heck 1991a) or individual data such as those listed in the directories published by the *American Astronomical Society (AAS)* (1990) or the *Société Française des Spécialistes d'Astronomie (SFSA)* (Egret 1991) should also be connected, as well as the electronic-mail data listed in Benn and Martin (1991) files.

Indeed one can imagine that someone searching the bibliography in SIMBAD can immediately get the communication 'plugs' of one author he/she wants to get in touch with. The same could apply with someone detaining the proprietary rights of some images from a given archive. Scientific profiles as listed in the ASpScROW and SFSA directories would offer other possibilities such as selective mailings for organizing meetings and so on.

Going one step further, one could consider the inclusion of dictionaries of acronyms and abbreviations (see e.g. Heck 1991b), as well as other tools that might come up as most useful in the everyday life (thesauri, S/W packages of various kinds, standardized desktop publishing macros or procedures possibly allowing direct connections to databases, and so on).

9. Conclusions or Where are We Going?

A general user does not really care in fact how the system he/she has at the tips of his/her fingers works, as long as it works well and give him/her as much as possible, and hopefully more than expected. This is why I did not enter here into technical details on databases structures, formats, protocols, and so on.

It is fair however to say that the user interfaces are also in permanent progress (see i.a. Pasian and Richmond 1991) allowing an ever more standardized access to heterogeneous sets of archives and databases. Hypertext and hypermedia systems have also been investigated as means for storage, retrieval and dissemination of information, for instance in the HST context (Adorf 1989).

Refer also to e.g. Adorf and Busch (1988) for uses of 'intelligent' text retrieval systems, to Kurtz (1991) for bibliographic retrieval through factor spaces, and to Parsaye et al. (1989) about 'intelligent' databases.

Of course, one is never happy and this is one of the major reasons behind the progress of sciences and techniques. Remember how we were only a few years ago. What an enormous progress in a short time. What will be the situation in only a decade from now?

Generally the evolution goes with quite a few surprizes and it would be stupid to play here the game of guessing what they will be. What is sure is that the future of databases and archives with be mainly linked to technological advances, but also to the fact that astronomers will more and more realize how intimately this field is involved with their activities: observing (remote or traditional), data reduction, general research, publication, management of instrument time, and so on.

Ideally, everyone would like to have from his/her desk (or home?) access quickly and in the most friendly way to the maximum of data. Upper shells should be accessible to interrogate databases and archives and retrieve the necessary material in a practically usable form. Local and cheap hard–, firm– and software should be available in order to then process this stuff locally.

Means of communications should eliminate the geographical separation. There might be a time when the keyboard will not be necessary anymore and when the publishing as such will disappear, replaced by some electronic form of dialog with the machine and the *knowledge* bases, a concept that will progressively replace those of databases and archives.

The present way we work is basically oriented towards the need of recognition (essentially through traditional publishing, even if eased thanks to the desktop tools) for getting positions (salaries), acceptance of proposals (data) and funding of projects (materialization of ideas). Most of the underlying rules were designed at a time when the pen (and then the typewriter) and noisy telephone lines over manned switchboards were the basic tools for communicating, with the index card as the access key to the archive (or library) when it was existing at all.

Human narcisism apart, is really recognition the best motivation behind the progress of science? How many redundancies could be avoided, as well as time and energy wasted in these and in administrative reporting of little fundamental use! The new era in archiving, in database interconnectability and in knowledge access, as well as in communication means, policies and habits might open the doors to new behaviours in our corporation.

Acknowledgements

My special gratitude goes to Miguel A. Albrecht and Daniel Egret for letting me have in advance of publication a copy of the book they edited on *Databases & On-line Data in Astronomy*. It is also a very pleasant duty to acknowledge here discussions with and/or material received from Hans-Martin Adorf, Jan Willem den Helder, Michael R. Garcia, Michael J. Kurtz and H. Ulrich Zimmermann.

References

Adorf, H.M. (1989): *Space Inf. Syst. Newsl.* **1**, 7.

Adorf, H.M. (1991): private communication.

Adord, H.M., Busch, E.K. (1988): in *Astronomy from Large Databases – Scientific Objectives and Methodological Approaches*, Eds. F. Murtgagh, A. Heck, *ESO Conf. & Workshop Proc.* **28** (ESO, Garching-bei-München), p. 143.

Albrecht, M.A. (1991): in *Databases & On-line Data in Astronomy*, Eds. M.A. Albrecht, D. Egret (Kluwer Acad. Publ., Dordrecht), in press.

Albrecht, M.A., Egret, D. (Eds.) (1991): *Databases & On-line Data in Astronomy* (Kluwer Acad. Publ., Dordrecht), in press.

American Astronomical Society (1990): *1991 Membership Directory* (American Astron. Soc., Washington).

Benn, Chr., Martin, R. (1991): *Electronic-Mail Directory 1991* (Roy. Greenwich Obs., Cambridge), e-mail files.

Burkhardt, G., Esser, U., Hefele, H., Heinrich, I., Hofmann, W., Krahn, D., Matas, V.R., Schmadel, L.D., Wielen, R., Zech, G. (Eds.) (1990a): *Astronomy & Astrophysics Abstracts* **49A**, Literature 1989, Part 1 (Springer-Verlag, Heildeberg).

Burkhardt, G., Esser, U., Hefele, H., Heinrich, I., Hofmann, W., Krahn, D., Matas, V.R., Schmadel, L.D., Wielen, R., Zech, G. (Eds.) (1990b): *Astronomy & Astrophysics Abstracts* **49B**, Literature 1989, Part 1, (Springer-Verlag, Heildeberg).

Burkhardt, G., Esser, U., Hefele, H., Heinrich, I., Hofmann, W., Krahn, D., Matas, V.R., Schmadel, L.D., Wielen, R., Zech, G. (Eds.) (1990c): *Astronomy & Astrophysics Abstracts* **50A**, Literature 1989, Part 2, (Springer-Verlag, Heildeberg).

Burkhardt, G., Esser, U., Hefele, H., Heinrich, I., Hofmann, W., Krahn, D., Matas, V.R., Schmadel, L.D., Wielen, R., Zech, G. (Eds.) (1990d): *Astronomy & Astrophysics Abstracts* **50B**, Literature 1989, Part 2, (Springer-Verlag, Heildeberg).

den Helder, J.W. (1991): private communication

Egret, D. (1991): *Annuaire de l'Astronomie Française 1991* (Soc. Française Spécialistes Astron., Paris), in press.

Egret, D., Wenger, M., Dubois, P. (1991): in *Databases & On-line Data in Astronomy*, Eds. M.A. Albrecht, D. Egret (Kluwer Acad. Publ., Dordrecht), in press.

Garcia, M.R. (1991): private communication.

Garcia, M.R., McSweeney, J.D., Karakashian, T., Thurman, J., Primini, F.A., Wilkes, B.J., Elvis, M. (1990): *The Einstein Observatory Database of HRI X-Ray Images (FITS/CD-ROM Version)* (Smithsonian Institution Astrophys. Observ., Cambridge).

Grøsbol, P. (1991): in *Databases & On-line Data in Astronomy*, Eds. M.A. Albrecht, D. Egret (Kluwer Acad. Publ., Dordrecht), in press.

Heck, A. (1986): *Bull. Inform. Strasbourg Data Center* **31**, 31.

Heck, A. (1991a): *Astronomy, Space Sciences and Related Organizations of the World – ASpScROW 1991, CDS Spec. Publ.* **16** (CDS, Strasbourg).

Heck, A. (1991b): *Acronyms & Abbreviations in Astronomy, Space Sciences & Related Fields, CDS Spec. Publ.* **18** (CDS, Strasbourg).

Heck, A., Murtagh, F. (Eds.) (1989): *Knowledge-Based Systems in Astronomy* (Springer-Verlag, Heidelberg).

Heck, A., Murtagh, F., Ponz, D.J. (1985): *Messenger* **41**, 22.

Helou, G., Madore, B.F., Schmitz, M., Bicay, M.D., Wu, X., Bennett, J. (1991): in *Databases & On-line Data in Astronomy*, Eds. M.A. Albrecht, D. Egret (Kluwer Acad. Publ., Dordrecht), in press.

Kurtz, M.J. (1991): in *On-line Astronomy Documentation and Literature*, Eds. F. Giovane, C. Pilachowski, *NASA Conf. Proc.*, in press.

Murtagh, F. (1988): *Bull. Inform. Strasbourg Data Center* **34**, 3.

Murtagh, F. (1991): in *Databases & On-line Data in Astronomy*, Eds. M.A. Albrecht, D. Egret (Kluwer Acad. Publ., Dordrecht), in press.

Murtagh, F., Heck, A. (1987): *Multivariate Data Analysis* (D. Reidel Publ. Co., Dordrecht).

Murtagh, F., Heck, A. (Eds.) (1988): *Astronomy from Large Databases – Scientific Objectives and Methodological Approaches, ESO Conf. & Workshop Proc.* **28** (ESO, Garching-bei-München).

Murtagh, F., Heck, A., di Gesù, V. (1987): *Messenger* **47**, 23.

Ochsenbein, F. (1991): in *Databases & On-line Data in Astronomy*, Eds. M.A. Albrecht, D. Egret (Kluwer Acad. Publ., Dordrecht), in press.

Parsaye, K., Chignell, M., Khoshafian, S., Wong, H. (1989): *Intelligent Databases* (J. Wiley & Sons, New York).

Pasian, F., Richmond, A. (1991): in *Databases & On-line Data in Astronomy*, Eds. M.A. Albrecht, D. Egret (Kluwer Acad. Publ., Dordrecht), in press.

Plummer, D., Schachter, J., Garcia, M., Elvis, M. (1991): *The Einstein Observatory IPC Slew Survey (FITSA3D/CD-ROM Version)* (Smithsonian Institution Astrophys. Obs., Cambridge).

Rolfe, E.J. (Ed.) (1983): *Statistical Methods in Astronomy, European Space Agency Spec. Publ.* **201** (ESA/ESTEC, Noordwijk).

Russo, G., Richmond, A., Albrecht, R. (1986): in *Data Analysis in Astronomy*, Eds. V. Di Gesù, L. Scarsi, P. Crane, J.H. Friedman, S. Levialdi (Plenum Press, New York), p. 193.

Schreier, E., Benvenuti, P., Pasian, F. (1991): in *Databases & On-line Data in Astronomy*, Eds. M.A. Albrecht, D. Egret (Kluwer Acad. Publ., Dordrecht), in press.

Turon, C., Arenou, F., Baylac, M.O., Boumghar, D., Crifo, F., Gómez, A., Marouard, M., Mekkas, M., Morin, D., Sellier, A. (1991): in *Databases & On-line Data in Astronomy*, Eds. M.A. Albrecht, D. Egret (Kluwer Acad. Publ., Dordrecht), in press.

Walker, H.J. (1991): in *Databases & On-line Data in Astronomy*, Eds. M.A. Albrecht, D. Egret (Kluwer Acad. Publ., Dordrecht), in press.
Wamsteker, W. (1991): in *Databases & On-line Data in Astronomy*, Eds. M.A. Albrecht, D. Egret (Kluwer Acad. Publ., Dordrecht), in press.
Watson, J.M. (1991): in *Databases & On-line Data in Astronomy*, Eds. M.A. Albrecht, D. Egret (Kluwer Acad. Publ., Dordrecht), in press.
Weiss J.R., Good, J.C. (1991): in *Databases & On-line Data in Astronomy*, Eds. M.A. Albrecht, D. Egret (Kluwer Acad. Publ., Dordrecht), in press.
Wells, D.C. (1991): in *Databases & On-line Data in Astronomy*, Eds. M.A. Albrecht, D. Egret (Kluwer Acad. Publ., Dordrecht), in press.
White, N.E., Giommi, P. (1991): in *Databases & On-line Data in Astronomy*, Eds. M.A. Albrecht, D. Egret (Kluwer Acad. Publ., Dordrecht), in press.
White, R.A., Mather, J.C. (1991): in *Databases & On-line Data in Astronomy*, Eds. M.A. Albrecht, D. Egret (Kluwer Acad. Publ., Dordrecht), in press.
Zimmermann, H.U., Harris, A.W. (1991): in *Databases & On-line Data in Astronomy*, Eds. M.A. Albrecht, D. Egret (Kluwer Acad. Publ., Dordrecht), in press.

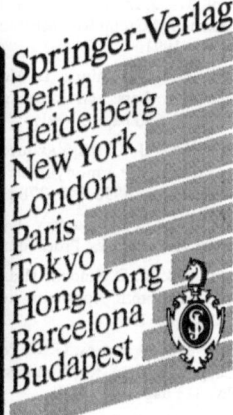

Lecture Notes in Physics

For information about Vols. 1–365
please contact your bookseller or Springer-Verlag

Vol. 366: M.-J. Goupil, J.-P Zahn (Eds.), Rotation and Mixing in Stellar Interiors. Proceedings, 1989. XIII, 183 pages. 1990.

Vol. 367: Y Osaki, H. Shibahashi (Eds.), Progress of Seismology of the Sun and Stars. Proceedings, 1989. XIII, 467 pages. 1990.

Vol. 368: L. Garrido (Ed.), Statistical Mechanics of Neural Networks. Proceedings. VI, 477 pages. 1990.

Vol. 369: A. Cassatella, R. Viotti (Eds.), Physics of Classical Novae. Proceedings, 1989. XII, 462 pages. 1990.

Vol. 370: H.-D. Doebner, J.-D. Hennig (Eds.), Quantum Groups. Proceedings, 1989. X, 434 pages. 1990.

Vol. 371: K.W. Morton (Ed.), Twelfth International Conference on Numerical Methods in Fluid Dynamics. Proceedings, 1990. XIV, 562 pages. 1990.

Vol. 372: F Dobran, Theory of Structured Multiphase Mixtures. IX, 223 pages. 1991.

Vol. 373: C. B. de Loore (Ed.), Late Stages of Stellar Evolution. Computational Methods in Astrophysical Hydrodynamics. Proceedings, 1989. VIII, 390 pages. 1991.

Vol. 374: L. Ting, R. Klein, Viscous Vortical Flows. V, 222 pages. 1991.

Vol. 375: C. Bartocci, U. Bruzzo, R. Cianci (Eds.), Differential Geometric Methods in Theoretical Physics. Proceedings, 1990. XIX, 401 pages. 1991.

Vol. 376: D. Berényi, G. Hock (Eds.), High-Energy lon-Atom Collisions. Proceedings, 1990. IX, 364 pages. 1991.

Vol. 377: W.J. Duschl, S.J. Wagner, M. Camenzind (Eds.), Variability of Active Galaxies. Proceedings, 1990. XII, 312 pages. 1991.

Vol. 378: C. Bendjaballah, 0. Hirota, S. Reynaud (Eds.), Quantum Aspects of Optical Communications. Proceedings 1990. VII, 389 pages. 1991.

Vol. 379: J. D. Hennig, W. Lucke, J. Tolar (Eds.), Differential Geometry, Group Representations, and Quantization. XI, 280 pages. 1991.

Vol. 380:1. Tuominen, D. Moss, G. Rudiger (Eds.), The Sun and Cool Stars: activity, magnetism, dynamos. Proceedings, 1990. X, 530 pages. 1991.

Vol. 381: J. Casas-Vazquez, D. Jou (Eds.), Rheological Modelling: Thermodynamical and Statistical Approaches. Proceedings, 1990. VII, 378 pages. 1991.

Vol. 382: V.V. Dodonov, V. 1. Man'ko (Eds.), Group Theoretical Methods in Physics. Proceedings, 1990. XVII, 601 pages. 1991.

Vol. 384: M. D. Smooke (Ed.), Reduced Kinetic Mechanisms and Asymptotic Approximations for Methane-Air Flames. V, 245 pages. 1991.

Vol. 385: A. Treves, G. C. Perola, L. Stella (Eds.), Iron Line Diagnostics in X-Ray Sources. Proceedings, Como, Italy 1990. IX, 312 pages. 1991.

Vol. 386: G. Pétré, A. Sanfeld (Eds.), Capillarity Today. Proceedings, Belgium 1990. XI, 384 pages. 1991.

New Series m: Monographs

Vol. m 1: H. Hora, Plasmas at High Temperature and Density. VIII, 442 pages. 1991.

Vol. m 2: P. Busch, P. J. Lahti, P. Michelstaedt, The Quantum Theory of Measurement. XIII, 165 pages. 1991.

Vol. m 3: A. Heck, J. M. Perdang (Eds.), Applying Fractals in Astronomy. IX, 210 pages. 1991.